Eminent Civil Engineers:
Their 20th Century Life and Times

edited by
David Doran

Whittles Publishing

Typeset by
Whittles Publishing Services

Published by
Whittles Publishing,
Roseleigh House,
Latheronwheel,
Caithness, KW5 6DW,
Scotland, UK

ISBN 1-870325-92-3

Printed by Redwood Books Ltd., Trowbridge

CONTENTS

Foreword

*I*t is entirely appropriate that the idea for this book should emerge from a discussion I had with Keith Whittles at 1 Great George Street. The corridors of this elegant London headquarters of the Institution of Civil Engineers are redolent with memories of eminent civil engineers. On the main staircase one's eye is arrested by such names as Telford, Rennie, Stephenson, Locke (but lacking a Brunel or two).

When we survey the landscape around us, we are continually reminded of the efforts and creative genius of engineers, whether in the form of a new road bridge, an opera house, a town centre development, an airport or perhaps more in the public focus, major projects such as the Channel Tunnel or the Millennium Dome. Perhaps we take the contribution of engineers to society rather for granted but, fortunately for us, there are those who commit themselves to this avenue of public service.

This book comprises a series of essays by some who have influenced civil engineering in recent times. Only history will reveal whether these engineers will be as revered as those in the past.

It has been my privilege to know or be acquainted with most of the contributors and an even greater honour to edit their work. Others were asked to contribute but declined for reasons of undue modesty or pressure of work. Our simple brief was to ask contributors to write about their careers – what set them on the road? – who has inspired and sustained them? – what have they found interesting? – how have they managed in a crisis? No clear pattern of career planning emerges: serendipity plays a major part.

The more observant of our readers may spot that one or two of our contributors are not civil engineers in the formal sense. I make no apology. Their contribution to the industry has been outstanding. Inevitably the pressure on space has meant that all contributors have had to be selective in what they have chosen to write. Some have elected to cover the whole of their career; others have chosen to write about a particular aspect of their work.

In so doing I hope they will encourage the young to enter a great profession with its endless fascination, challenge and intellectual reward. Equally this book will appeal to those well-established in the industry and those to whom it is of general interest.

I would like to thank all the contributors; Keith Whittles, my commissioning editor and publisher; Peter Campbell for his help in selecting a list of potential contributors, and finally my wife Maureen for her help and encouragement.

David Doran

Introduction

David Doran

A romance with engineering

My romance with engineering began when I was quite young. An older foster brother, later to become chairman of a major construction company and to be knighted, told me stories of building sites and the personalities that ran them. Later, at the age of seven, I contracted rheumatic fever and an observant doctor encouraged my architectural sketching during my convalescence.

A predilection for dismantling family heirloom, pre-quartz watches suggested a future in mechanical rather than civil engineering!

Choice of secondary education heralded a mild rebellion from the norm. Having passed the 'eleven plus' examination I opted for the local technical college rather than grammar school. The South West Essex Technical College was interesting: you could enrol at eleven and then emerge ten years later with a university degree! The early 1940s were dark days with more time spent in the air-raid shelter than in the classroom: truly an education at the university of life. Most of us did not matriculate until 17 so we had a lot of catching up to do. For recent school leavers like myself, degree courses were galvanized by those returning from the armed forces, many of whom already had HNC or HND qualifications. Unpleasant arguments arose in the common-room between older and younger factions.

The standard of instruction during my degree course was not inspiring although there was much emphasis on the basics of drawing, particularly sketching, which developed powers of observation and a recognition of proportion. Mathematics was well taught and most of us were at MSc standard by the time that finals came around. An external London University degree (two years from inter-BSc) was hard work: no exam mark allowance for course work and little clue as to which part of the syllabus would be examined. We were assured that Jansen's theory for silos would not appear again but regrettably Murphy's law (anything that can go wrong will go wrong) applied.

My introduction to a civil engineering site was as a vacation student at the West India Dock (see Figure). Repairs to war-time bomb damage were under

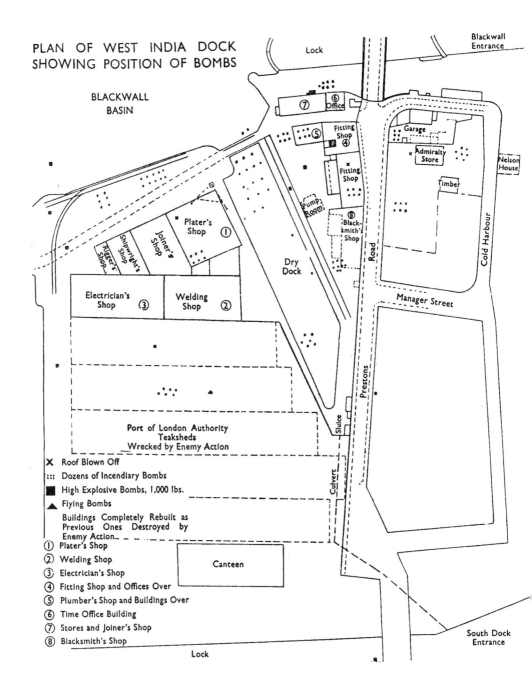

PLAN OF WEST INDIA DOCK
SHOWING POSITION OF BOMBS

BLACKWALL
BASIN

Lock

Blackwall
Entrance

⑦　⑥ Office

Fitting
Shop
⑤　④

Garage

Admiralty
Store

Nelson
House

Fitting
Shop

Timber

Pump
Room

Plater's
Shop　①

⑧
Black-
smith's
Shop

Road

Cold Harbour

Joiner's
Shop

Rigger's
Shop

Shipwright's
Shop

Dry
Dock

Electrician's
Shop　③

Welding
Shop　②

Manager Street

Prestons

Sluice

Port of London Authority
Teaksheds
Wrecked by Enemy Action

Culvert

X　Roof Blown Off

:::　Dozens of Incendiary Bombs

■　High Explosive Bombs, 1,000 lbs.

▲　Flying Bombs

Buildings Completely Rebuilt as
Previous Ones Destroyed by
Enemy Action

① Plater's Shop
② Welding Shop
③ Electrician's Shop
④ Fitting Shop and Offices Over
⑤ Plumber's Shop and Buildings Over
⑥ Time Office Building
⑦ Stores and Joiner's Shop
⑧ Blacksmith's Shop

Canteen

South Dock
Entrance

Lock

2

way. The experience was somewhat daunting when compared to the cosseted environment of home and college. One day I ventured across the top waling of the cofferdam before the guard rails had been erected. On one side were the murky waters of the Blackwall Basin, on the other a 25-foot drop to the base of the dry dock. Overcome by fright I turned back. Confronted by the smiling face of a burly Irish foreman, I was encouraged to continue. Somehow I made it.

Little did I realize at the time but my future father-in-law, the late Captain Robert Massam, was working as dockmaster of the Regents Canal Dock only a few hundred yards away. I have at least two things to thank him for, firstly my dear wife Maureen and secondly a bequeathed copy of the 1920's civil engineering classic *Dock Engineering* by Brysson Cunningham.

Getting a job after graduation was easy and I soon found myself at Wimpey amongst such luminaries as Sir Godfrey Mitchell, Dr William MacGregor, Dr Leonard Murdock, Michael Tomlinson and many others. As an indentured engineer 'Dr Mac' was my supervising engineer, although I spent much of my early time with Len Murdock. In my first encounter with him he said: 'Now Doran, I know you have an honours degree, forget about that, your real education starts today.' A deflating introduction not helped by a first week's pay packet of £4 19*s* 11*d*! The subsequent experience was, however, excellent as my first two years gave me site, design and laboratory experience and a bursary to attend a postgraduate course at Imperial College, London.

Site experience at Ellesmere Port was enlivened by frequent strikes called by Eric Heffer, then a strong union man but soon to become a respected left-wing Member of Parliament.

Early in my working career I was required to supervise two Irish drilling gangs on a major site investigation. I learnt something about soil mechanics but a great deal more about life. Each week I collected the team's wages in cash from the local post office. These were distributed on Thursday and by the following Tuesday they were usually spent and I then received requests for a 'sub'. After New Year's Eve two gangers appeared with obvious facial damage from a fight at the pub. Inter-gang cooperation was not too good in the first few days in 1951. Such were the rich rewards of life on site.

Our professor of concrete technology at Imperial College was Arthur L.L. Baker, affectionately known as A.L. 'squared' Baker. He was supposed to lecture us on concrete design. In practice, however, after the first ten minutes of any class he usually stepped up a gear and regaled us with concepts for crossing the English Channel and bell fendering systems. He also recommended the use of a Chinese abacus to solve simultaneous equations in shell roof design. Imperial introduced me to Professors Skempton and Pippard – an experience not to be missed – each pre-eminent in their specialities of soil mechanics and structures respectively. For me this course raised my interest in conceptual matters and engineering in the round. Many years later it was my pleasure to be awarded a Fellowship of the City and Guilds Institute, an organisation closely associated with Imperial College.

National Service still existed and after several years of deferment I was required to join HM forces. Rejected by the RAF airfield construction branch because I was unwilling to sign on for three years, I found myself in the Royal Engineers. I was posted to Taiping in North Malaya and the realities of a country at war; not a conventional war but one where Communist terrorists acted under cover of impenetrable jungle. Travel in some areas was restricted; in 'grey' districts you travelled under escort and in convoy in 'black' districts. Imagine my surprise when, long after demob, I discovered that my favoured driver, a Malay corporal, had been arrested as a suspected Communist sympathizer.

I was well treated by Lt. Col. L.T.C. Rolt the CRE*, and the civilian ACRE* Bob Northey, a giant of a man and a qualified structural engineer. I designed (and saw constructed) barrack buildings and bridges and helped pioneer a method of industrialized construction of billet blocks using steel Braithwaite tank panels as concrete formwork. The occasional sortie into the jungle to build a Bailey bridge for the PWD* was rewarding, particularly on one occasion when we discovered that the Hussar armed detachment detailed to protect us from terrorists had only been in Malaya for ten days.

Resuming at Wimpey in 1955, I was surprised how much faith people had in me. Soon I became project design engineer for surface works at Parkside Colliery, Newton-le-Willows. Sir Godfrey, who was at that time a non-executive director of the National Coal Board, had responded to the challenge of reducing the time from finding mineable coal to the production of the first saleable ton. We helped reduce this period by five years. I soon became familiar with the quaint terminology of pit-head structures – fan drifts, evasées, clarifloculators, run-of-mine bunkers and washery plants. Sadly the pit, which was commissioned in the early 1960s, is now closed.

The construction industry is a hotbed for disputes. Many of these arise because construction has been undertaken before completion of working drawings. In many years in the industry, Parkside was the only job I can recall where all the details were completed before construction was commenced. The job was a success, there were no disputes, the client was pleased and the designer/contractor made a profit.

Other regular clients at this time were Metro-Vickers, Trafford Park, Manchester and the Clydeside shipbuilders John Brown, builders of the great Queen liners. JBs once requested the design of the foundation to a vertical drilling machine to have 'zero settlement'. A tall order even for Professor Skempton.

Wimpey was a benevolent, family-type company and one that I enjoyed. It was full of characters, some charming, some not so, and others with a harshness borne of the exigencies of running difficult jobs on remote sites. When, at 36, I eventually became a manager I was given one particularly good piece of advice by E.G. Loudoun, a main board director and a chartered structural engineer. He said: 'For one day a week lock your office door, take the 'phone off the hook, sit behind your desk and think' – good advice but difficult to put into practice.

*CRE – Chief Royal Engineer; ACRE – Assistant Chief Royal Engineer; PWD – Public Works Department

I was fortunate to become manager of the structural design department and subsequently the civil and structural design department responsible (at peak) for 250 staff. Directorships of three subsidiary companies followed. This broad experience taught me many things including :

- You can delegate everything except the responsibility.
- How to develop a 'grasshopper' mind.
- How to control a check list with rarely less than twenty items awaiting attention. Stress levels can be greatly reduced by dealing with ten 'five minute' items before returning to more weighty matters.
- That in any human society things will go wrong. (Murphy understood that!)
- Patience and perseverance.
- That human beings are unique and that blanket and uniform approaches do not work with individuals.
- To adopt an 'Eisenhower' rather than a 'Montgomery' approach to life. (In WW2   General Eisenhower favoured working on a broad front rather than the narrow thrusts favoured by Field Marshal Montgomery).
- As the Bible says, live a day at a time. (Matthew 6:34).
- To do the best one can in a short time, often with limited facts and resources.

Dudley Dennington was my predecessor. I recall a discussion in which I said, 'My formula for success in professional life is a mixture of ability and opportunity'. He disagreed and offered 'a mixture of good luck and good health'. I wonder who was right?

Wimpey prospered at home and abroad. One of my most taxing tasks was to deal with rapidly deteriorating reinforced concrete structures in the Middle East. At the relevant time the method used for winning aggregates was for a family with a truck to drive into the desert and load their vehicle with material crow-barred from the surface. This stone was then crushed, stockpiled and subsequently sold to contractors. In the 1960s it was not generally realized that the material thus obtained was heavily laden with chlorides which eventually led to the rapid corrosion of steel reinforcement. One structure constructed using this aggregate became, for a while, a national monument appearing on postage stamps, calendars, etc. Soon after completion it cracked up badly and I acutely remember a meeting with the government chief engineer who said to me 'Mr Doran, we did not pay your company to build a cracked structure and would like it returned to its uncracked state.' This was one of the more demanding meetings of my career. The profession today is much more knowledgeable but considerable time lags still occur between the understanding of a problem and the adoption, in practice, of a solution. Perhaps expert systems will help to close this gap.

At a later date I visited Romania with a colleague to assist in remedial work following a severe earthquake. This was just one of several visits behind the 'Iron Curtain' to countries such as Russia and Bulgaria. Other interesting ventures were to apartheid-ridden Johannesburg and to Belfast to pick up the pieces after a terrorist attack on the *Daily Mirror* printing works. The latter involved inadvertently walking over unexploded bombs hidden beneath rubble! An even more rewarding experience was to be associated with the construction of the 47-storey Hongkong & Shanghai Bank, an award-winning design by Norman Foster.

In the 1970s, with encouragement from Dr. Jack Chapman (then technical director) I began to take a greater interest in professional institution affairs. Election to the Council of the Institution of Structural Engineers followed and I eventually became honorary secretary of that Institution. The secretariat and others recruited me to become chairman of various technical committees and task groups. As a result, I became involved with topics as wide ranging as alkali-silica reaction in concrete; detailing of reinforced and prestressed concrete; appraisal of existing structures; bridge access gantries; cladding and the management of design offices. In 1993 I was given the Lewis Kent Award for my services; an unexpected but highly appreciated honour. I have to thank a broad experience at Wimpey to have enabled me to play a part in these varied tasks.

In 1985 there was a sea change in my career. I left Wimpey and became an independent consultant. From a big, all-pervading umbrella to a very small one. I had to rebuild my confidence and motivation. Many rewarding experiences in contracting and the army gave me a springboard from which to launch a second career. I soon found myself approached by other consultants to be retained on their staff. In such a capacity I have worked with Freeman Fox & Partners, Bates & Gillibrand, and Technotrade Ltd.

In all of this I have been loyally supported by my wife, Maureen, who has also aacted as my business administrator. It has also given me a chance to put something back into a profession from which I have had so much enjoyment.

In the late 1980s I was fortunate to become involved in publishing and have been responsible for editing books on construction materials with Butterworth Heinemann and Thomas Telford. These have been exercises in project management in attempting to guide many expert authors along a predetermined path. The latter part of my career has also become enriched by becoming an advisor to Dr Keith Whittles, the publisher of this book.

I have always had an interest in the education and training of engineers and have regarded the true professional as one who keeps up-to-date. This updating procedure is now known as continuing professional development. In today's busy world this process is assisted by distance learning techniques including the use of videos and study packs. As part of my continuing career I now advise the Television Education Network in these initiatives.

So, in conclusion, life in civil engineering has spanned the years and the emotions – it has been demanding, rewarding, stimulating, occasionally frustrating but never dull. If not an engineer, what might I have been? Perhaps golf correspondent of the *Daily Telegraph* or a jazz pianist like Oscar Peterson. Not a bit of it – give me civil engineering every time.

When you have read the other contributions in this book I hope you will agree with me.

DAVID DORAN

BSc (Eng), DIC, FGGI, CEng, EurIng, FICE, FIStructE, MSocIS(France)

David Doran spent his early career on site, gaining construction experience of process plants, jetties and a large steel-framed office block with basement. After a year with Wimpey Laboratories and a further postgraduate year at Imperial College studying concrete technology, he spent his National Service with the Royal Engineers, mainly in Malaya.

He took early retirement after many years with George Wimpey. For the last 20 years he was chief structural engineer and more recently a director of Wimpey Group Services; Wimpey Construction UK and Wimpey Laboratories. He was responsible for quality assurance in Wimpey Construction UK and Wimpey Homes Holdings. He was chairman of WCUK/WLL research liaison committee, a member of the technology board and advised the Wimpey Group on technical litigation.

Since July 1985 he has been practising as a consulting engineer in the following fields: advisory work, structural surveys, arbitration, quality assurance, technical audits, litigation, and insurance claims.

His spectrum of permanent and temporary works design at home and abroad includes housing (traditional, no-fines and timber-framed), light industrial and process plants, commercial and property developments, marine and civil engineering, and offshore.

He was a member of Council of the Institution of Structural Engineers for nine years in the 1980s and has served on a number of technical committees at ICE, IStructE, BSI, IABSE, Concrete Society and Imperial College. He has also been extensively involved in publishing and with distance learning techniques.

Coordinating expertise to meet new challenges

David Quinion

*W*hen in 1946 I left university to commence work as a contractor's engineer under an indentureship it was not possible to plan the development of a career beyond achieving membership of the ICE and aiming to reach the status of a site agent. Construction was a competitive industry in all respects. Engineers, like other employees, were subject to hire and fire; the hours were long. Contracts needed to produce good results, many essential materials were in short supply and training was very much learning by experience. The experience lay more with general and trades foremen than with site management staff. Working with such men, one had to strike an accord and find out from them the reasons governing the construction methods used and preferred. This was necessary because there was little available in print, and mostly it concerned methods of design which needed to be interpreted for a temporary works situation or for sorting out problems arising on site. Overcoming the supply shortages and delays of materials and improvisation provided challenges which had not been anticipated at university.

In 1955 another step change presented itself with the opportunity to form a team and develop with engineering teams of other disciplines a package deal proposal for a first commercial nuclear power station. Coordination and management of such a team involved different approaches to one's own staff and to those of the other teams to make the best use of their knowledge and ideas.

In 1974, while employed as company chief engineer of Tarmac Construction, I had the opportunity to write a draft code of practice for falsework. I shall describe why I accepted that challenge and how it was met by searching out the available expertise across the construction industry and effecting an understanding and endorsement of the final BS5975.

Castle Donington power station in the Trent Valley

I was recruited by Taylor Woodrow Construction in 1952 as the site Chief Engineer for the foundation contract (their first £1 million contract), and we were subsequently awarded most of the other building and civil engineering work on what was the largest European power station at the time.

The power station was close to the river Trent and required most excavations to a depth of 5.1 m through sand and gravel down to Keuper Marl. The site was protected from the river by an earth bank but the river could rise by some 3 m in 36 hours and frequently flooded the valley beyond the river. The agent and many of the foremen had worked together before, and with managing director Tom Reeves back to his days on site, but the engineering staff were new to each other. The foremen were highly experienced and I made use of their ideas.

The first major challenge was the main excavation for the foundations to boilers and turbines. The boilers were to be hung from the structural steelwork and the foundations were to be made of colloidal concrete in large pours. The foundation for each turbine block would be poured in one continuous operation 24 m × 12 m × 5.1 m. It was decided to open up one large excavation with an access ramp down to formation level remote from the river. The site investigation indicated high permeability of the granular soils. The excavation would have the sides supported by anchored sheet piling to exclude flows of groundwater. There was time before the award of the contract to examine the temporary works materials in the plant yard and we found an abundance of former excavator wire ropes and a miscellany of steel sections. So I designed the sheet piled walls with these materials to provide walings, tie backs and old short piles as anchors. There were long discussions as to how much the ropes would stretch and how to make the best of this cheap solution. It worked well and allowed the foremen to develop confidence in their engineers. After high early inflows through the unpiled open end we kept the groundwater lowered with minimal pumping.

Colloidal concrete had been developed to eliminate cracking on large pours. The shuttering would be filled with 125 to 60 mm stone pieces packed in by hand and later grouted. The theory was that, since all the stones were packed in and touching, then after grouting there could not be any shrinkage. So we required a 5.1 m shutter around each block which would be slowly grouted over nearly 1½ days using a heavily retarded and workable grout. We developed our own design criteria for the shuttering. In the pour were standpipes for grout injection and monitoring, large circulating water pipes, anchorages for later post-tensioning of the TA (turbo-alternator) blocks above and large holding-down bolt assemblies. The work was successful, although despite the stone packing there were movements as the grouting operations lubricated the contacts between the stones.

The TA blocks were highly complex arrays of concrete columns, beams and slabs to be poured each in one pour using high strength concrete of specified 40 mm maximum slump. They were heavily reinforced and had also to be post-tensioned. There appeared to be no spaces greater than 200 mm cube and the overall height was about 7 m. Although we recognized the challenge and prepared well, the concreting was too complex despite providing openings for insertion of poker vibrators and then an easement of the slump. We had a long examination of, and then pressure grouting to make good, No. 1 block, and then a joint exercise with the designers to alter the reinforcing to provide more reasonable access for concreting. There were no further problems but I learnt to anticipate when concreting results could not be assured and that designers and contractors can resolve such problems together. Some years later, when superplasticizers became available from overseas, I researched them for CIRIA (Construction Industry Research & Information Association) and a guide was published concerning their use.

One night disaster struck when the Trent overtopped and burst its banks, flooding the cooling water pumphouse and system. It made the newspapers, disturbed head office, client and consultant and I happened to be acting agent. I rang the managing director, who said he would keep off-site worriers at bay and recommended approaching the Fire Brigade to pump out the water as being quicker than trying to bring in more usual pumping equipment. He gave us four days to get it dry. The Fire Brigade enjoyed the change and the extra money, and we met the deadline. By now the site team had complete confidence in each other and our joint abilities.

We employed the Hoyer long-line prestressing systems to make precast roof units for culverts; we used soil stabilization for site roads, undertook one of the first applications of vibroflotation stone columns for foundations to the cooling towers, had two weekend occupations of the main railway line whilst installing culverts underneath it and used other techniques which were novel or challenging.

I had developed team working between engineers and foremen on the site so that we attained confidence in our collective activities; we had a good measure of those abilities and knew how to interchange ideas with the designers for whom we worked. It prepared me for my next quite unexpected challenge.

Design and construction of a nuclear power station

The commercial nuclear power station programme was launched with four competing consortia. One comprised English Electric, Babcock & Wilcox and Taylor Woodrow, and I was appointed chief civil engineer. Initially a small team of 18 started on the design of a reactor with no preconceived ideas on the configuration of the pile, and started to assess the alternative options with the objective of submitting tenders in 18 months' time. Every decision considered technical performance, likely cost and effect on programme. Many major decisions involved reconciling different viewpoints. It was vital to have confidence and trust not only in one's own staff but in those in the other two teams and to convey that confidence to the company's directors. It meant having confidence in one's own abilities to assess data, to draw the right conclusions and convince others without stifling ideas. Often instinct would lead one to look for further data before reaching a decision. As the design developed it was monitored by Sir Christopher Hinton and staff from the CEGB as the clients and the UKAEA as their consultants.

Three sites were selected for tenders and a joint working party from the consortium agreed on site investigations sufficient for their needs without disclosure of the individual station layouts. Our layout reflected our views as contractors to achieve the best construction programme and use of resources. My design team incorporated Allotts to design the turbine house and cooling water system, Frederick

11

Gibberd as architect, United Steel for steelwork, Hadens for mechanical services and Troughton & Young for electrical services. My staff undertook the design of the reactors, services, buildings, roads and external services. In appointing staff it was important to get expertise, with enthusiasm and a team attitude. It was also necessary to manage the flow of ideas and the decision making without too much restraint. Each of the consortium members nominated a leader to whom documents were addressed and who received copies of all outgoing ones. These were assessed by a senior engineer who arranged for immediate circulation of copies,

The Goliath crane straddling the reactors (courtesy of BNFL Magnox Generation).

indicated who would be responsible for coordinating any action and then all those of possible importance were passed to me. This worked very well and kept everyone informed as to progress which could affect their own work and also prevented items being overlooked. It enabled me to assess progress and identify difficulties. We also operated extent of supply schedules which detailed who was responsible for every item to be provided to satisfy the consortium that nothing would be omitted or duplicated in the tender.

We were not successful on the first three stations but had a further nine months to tender for Hinkley Point. In this time we were able to increase the output to 500 MW from 320 MW by further optimization and critically reviewing many decisions. The Hinkley Point site was quite different to the others. We were able to locate the turbine house and CW pumphouse on an area of foreshore which we reclaimed from 100 years of cliff erosion by the construction of a sea wall in tidal conditions. This saved on excavation and pumping head. The reactors governed the overall programme for the station and a novel approach was adopted to avoid building the concrete biological shield walls and the reactor pressure vessel in sequence.

The shield walls would be built whilst rings of the spherical steel pressure vessel were fabricated in a temporary building alongside. The rings of the pressure vessel and the shells of the heat exchangers would then be lifted into position by a Goliath crane (see Figure) with a span of 76.2 m straddling the two reactors and the fabrication building. It could lift 400 tons to a height of 61 m and was the largest crane of its type in the world. Its enormous cost was justified by savings in time and alternative erection plant.

First the biological shield walls had to be designed and we could find no comparable design elsewhere. The 2.4 m thick shield walls would be subjected to direct heat from the reactor on their inner faces and heated internally under the bombardment of the neutron radiation it was designed to absorb. These together gave rise to complex heating patterns over the height of the walls, and months of finite element analysis were invoked before we arrived at a satisfactory method of analysis and detailing. There were many other problems to be resolved in the reactor area for which there were no established precedents, and a major one was the pilecap through which load and discharge operations took place.

During the construction period I became responsible in Taylor Woodrow for the implementation of the tender. I oversaw the execution of the design, dealt with the contractual implications with the CEGB and our consortium partners, coordinated overall progress of construction and design, and had a quality assurance team on site.

Throughout the pre-tender and post-tender periods the need was to encourage the interaction of different disciplines and recognize their priorities whilst maximizing the benefits of undertaking design and construction. Success was dependent on design and site staff anticipating their future activities and cooperating with each other. This they did, so that the work was undertaken within the tender sums, to programme and without recourse to contingency sums.

13

Contributing to codes and standards

After nuclear power I had another major design and construct project in the rebuild of Euston Station, when again my staff were largely technically self-sufficient. When I was appointed the company chief engineer to Tarmac Construction I was concerned with different types of projects, most of which were too small to sustain many technical staff and relied upon a central engineering department to provide them with construction methods, designs of temporary works and general technical planning support. As the workload increased, I recognized advantages in creating specialisms within that department, setting standards for what was produced, providing a trouble-shooting service and benefiting from feedback. Necessarily I wondered whether our standards were as good and as competitive as we believed them to be, when many competitors appeared to economize and get away with lower performance standards. Involvement in the technical affairs of the Institutions, Concrete Society and CIRIA helped to assess what other companies did and to identify general concerns for which guidance would be welcomed.

Involvement with standards-making committees and guides to good practice gave access to the current thinking about such codes, how they should be applied and their limitations. This helped in resolving problems arising on site or on completed projects where the applications of such standards and codes were called into question. It was clearly beneficial to acquire such knowledge. Following worldwide concern about the high frequency of collapses of falsework, a committee under Professor Bragg made recommendations to rectify the situation and a BSI Committee was appointed to prepare a code of practice. They decided to appoint a consultant to prepare a draft code for them and I obtained the appointment. There were no existing codes of practice for temporary works and little authoritative advice available worldwide, so it was a challenge, with myself and Tarmac becoming targets for criticism. Falsework was being undertaken by many types of organizations and varied from supporting major bridge decks to cast in-situ concrete beams.

To learn how other companies, general and specialist, undertook falsework design and set their standards of performance, we surveyed over 50 sites and • visited some 50 organizations. Apart from wishing to learn the basis of current design, we also wished to learn how designers of the permanent works checked and passed the designs of falsework submitted to them. In the ninth month it was necessary to produce the draft, but much of the evidence was inconsistent and more factual than based on reasoning or test work. Current good practices seemed quite adequate and in part could be substantiated by reference to permanent works design methods and applying 'temporary use' factors. I decided the code should be a statement of current good practice. The draft code was written in handbook style to:

   a) Explain the investigation behind the recommendations.
   b) Give design information in tabular form for simpler falsework applications.

c) Provide design methods and lists of factors to be taken into account.

d) Provide a self-contained code with extracts from other codes and standards and advice on using them.

e) Explain the approach to design for wind loads and assessment of ground conditions.

f) Recommend proper care and identification of materials used in falsework.

g) List common causes of failure and measures to be taken to avoid them.

h) Advocate the appointment of a falsework coordinator for each project to ensure the systems recommended for design, erection, use and dismantling are applied.

A long period of consultation, explanatory meetings and committee discussions led to the Code for Falsework BS5975, which contained all of the above with less of the explanation as that was no longer deemed necessary.

This code set the standard for falsework as that of current good practice and required all practitioners to conform, so reducing the incidence of low standards, low prices and failures. The construction industry was able to apply similar approaches to other types of temporary works and produced guides to good practice for most other construction methods. Designers of permanent works had guidance to the principles used by contractors and a reference against which they could assess the adequacy of designs submitted to them. Being involved in the falsework code and many other guides to good practice enabled my staff and I to be confident that our standards of performance were safe and competitive and encouraged confidence by site staff in our designs.

Concluding remarks

The construction industry requires the coordination of clients, consultants, contractors, subcontractors and suppliers to assuredly provide appropriately designed and constructed works based on recognized good practices.

It is important that the construction industry sets out unambiguous standards for its performances and that these reflect the views and expertise of the various parties involved. Those who design and those who construct have their different contributions to definitions of good practice and should understand and have confidence in them. Other engineering disciplines can provide both ideas and criticisms.

My experience at Castle Donington taught me how to coordinate the practical experience of the foremen with the knowledge of the engineers. Experience on nuclear power station development design and construction taught how to coordinate multi-disciplinary experience, engineering, design and construction practices on challenging major projects. The scale of coordination widened still further in drawing on the expertise of the construction industry in writing a code of practice for falsework and other guides to good practice.

David Quinion

BSc (Hons Eng), FREng, FICE, FIStructE

Born 1926.   Graduated: Northampton Engineering College.

1946–52 Site experience: Sir R. McAlpine & Marples Ridgway & Partners (also estimating); design experience: Oscar Faber.

1952–67 Taylor Woodrow Construction: chief engineer on Castle Donington Power Station; chief engineer, development and design of nuclear power stations and for implementation of designs/profitability of Hinkley Point and Sizewell, contracts manager for Euston Station and other projects.

1967–88 Tarmac Construction: company chief engineer responsible for technical performance, provision of central engineering services, geotechnical division, quality management, technical developments and problem solving. Involved in many national and international activities concerning technical standards.

Go measure earth

Michael Tomlinson

> *Go wond'rous creature! mount where Science guides,*
> *Go, measure earth, weigh air, and state the tides!*

Pope, *Essay on man*

*A*lexander Pope's command to measure earth and state the tides is an apt summary of my career in civil engineering, which has been in most part concerned with earthworks and foundations, including considerable maritime construction. However, when I commenced work in 1934 as an articled pupil to a borough engineer, my ambitions were very limited and I had little idea of the wide range of activities which can be enjoyed in the working life of a civil engineer.

In the 1920s and 30s a frequently-used route to corporate membership of the Institution of Civil Engineers was to become an articled pupil to a chartered civil engineer. The engineer was usually a busy man who had little time available to teach his pupil. The student had to learn what he could from older colleagues and by evening study at a local technical college. If no suitable courses were available locally the student would take a correspondence course – the worst form of engineering education. The 'tuition' fees would have to be paid by the parents.

I was in this unfortunate position when after leaving school in 1933 I became articled to the engineer and surveyor to the Scunthorpe and Frodingham Urban District Council (soon to become Scunthorpe Borough Council). Courses in engineering at the local technical college were confined to the technology of the iron and steel industry. Consequently there was no alternative to studying for the Associate Membership examination through a correspondence course. However, I was fortunate in my practical training because our small office was busily engaged on quite sizeable improvements to the town's water supply, sewerage and highways, consequent on industrial expansion in the area.

I had had no boyhood longing to become a civil engineer but upbringing must have had some influence on my eventual choice. My earliest years were spent in sight and sound of the steelworks with the flaring blast furnaces and nightly sounds of shunting engines and wagons hauling iron ore and coal. I had been pushed in my pram past the opencast mine workings where labourers were still stripping the overburden by hand and wheeling their heavy barrows across trestle planks high above the rail wagons waiting below to be filled by the steam navvy excavating the ironstone.

In the first two or three years of my pupillage I had little idea as to where my career might lead. Probably my ambition extended no further than eventually becoming a borough engineer in some northern town – but this was not a lowly ambition. In past years this title had considerable dignity and status before the duties and responsibilities of local authority engineers were savagely emasculated by the governments of the 1980s. In fact, these responsibilities have decreased in recent years in inverse ratio to the grandeur of their titles, the borough engineer becoming 'the director of the built environment'. However, in 1936 an event occurred which had a profound influence on my subsequent career and my out-look on civil engineering. In that year the firm of Sir Robert MacAlpine & Sons was awarded the contract to build a new ironworks for the Appleby-Frodingham Iron and Steel Company. This was a major project for increasing steel production as a forerunner to the nation's rearmament programme. At the time MacAlpine was one of only two or three contractors in the country with the resources and experience to do the work in the short time available.

Our small office had no facilities for plan printing and once or twice a week, as the most junior member of staff, I had to take rolls of tracings in my bicycle basket

The foundations of one of two blast furnaces and associated plant (Author's photograph).

to the steel company's office for printing. While waiting for this to be done I would visit the construction site of the new ironworks. This might be considered as wasting my employer's time but certainly I was not wasting my own. I was able to watch the progress of many operations such as large-scale earthmoving, construction of blast furnace and blowing engine foundations, ore-stocking yards, bridges, and miles of rail sidings.

The site was covered by about ten metres of slag fill, silt and sand overlying rock. Piling for the foundations was not feasible because of massive obstructions in the fill. In the pre-war years heavy equipment for sinking large diameter bored piles had not been developed. Instead the whole area was excavated down to rockhead, which involved the excavation by face shovel and dragline of 2.5 to 3 million cubic metres of material. The foundations were then constructed and the excavated spoil brought back by steam-hauled standard-gauge tipping wagons and backfilled by draglines around the reinforced concrete substructures as shown in the accompanying photograph. The buttressed retaining walls of the ore-stocking yard were concreted in two 6 m lifts surmounted by the 2.4 m deep rail beams carrying ore-handling equipment. The heavy timber formwork panels were lifted and placed by 15 tonne travelling steam derrick cranes. Concrete was conveyed from the mixing plant on rail wagons.

During these snatched visits I was able to wander everywhere over the site. If challenged by anyone in authority I would say that I was from the borough engineer's office giving them the impression – quite wrongly – that I had some official status on the job. I got to know MacAlpine's young engineers and foremen and would meet them from time to time in the local pubs and on the rugby field. Two of these engineers had just returned from building grain silos in the Argentine. This widened the outlook of a rather green youth who had rarely travelled far from his native Lincolnshire. Such inspiration could never have come from the brief red-ink comments from the anonymous tutors on my returned correspondence college course work.

I watched with amazement the nonchalance of Laird Adams, MacAlpine's chief agent (who nowadays would be called the 'director of construction'), in his shirtsleeves and wearing plus-fours, as he personally supervised the offloading of a steam locomotive brought in by low-loader from some distant contract. The vehicle was manoeuvred across the railtracks at a level crossing on a main road through the steelworks. This operation held up the road traffic for hours: an example of contractors' philosophy at that time of 'Do it first and ask permission afterwards!'

The blowing engines for the blast furnaces were manufactured by MAN of Duisberg. A team of German erectors came to install them. One of their number soon established an Anglo-German Fellowship Bund in the town with the object of gaining sympathy for the Nazi ideals. Although the Luftwaffe must have been well aware of the extent of the steel industry in the Scunthorpe area, the works were virtually undamaged from bombing in the 1939–45 war. This was probably because Hitler hoped for an intact steel industry in Britain as part of the victory spoils.

19

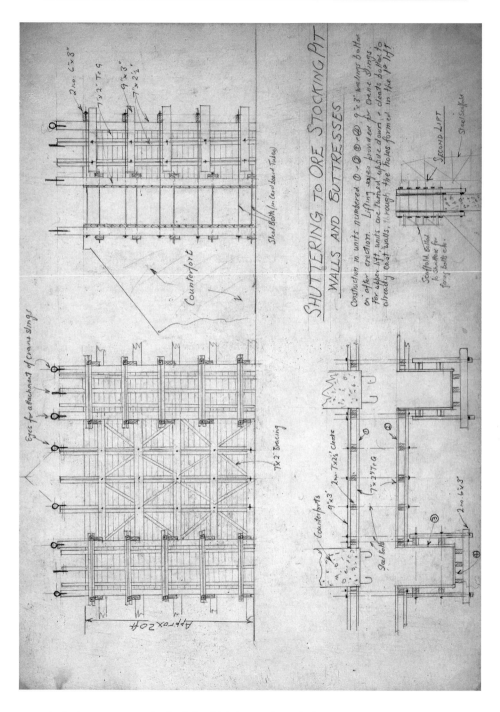

Author's contemporary sketch of 6 m high formwork panels for ore-stocking pit walls.

As a means of absorbing details of construction methods I learned the value of freehand sketching. I was becoming increasingly attracted to the life of a contractor's engineer but I realized that I would need to know about the design of such things as the huge formwork panels and timbering of deep excavations in use on the site. So on summer evenings I would bicycle with notebook and pocket tape to sketch and measure the size and spacing of the principal structural members. A camera was used only sparingly to record the general proportions for later preparation of the finished drawings. In the following years I continued to take opportunities of sketching construction work. This gave me an instinctive feeling for the appropriate size of structural elements, and on more than one occasion this has saved me from embarrassing mistakes in calculations; I have always been a poor mathematician.

I do not like to see young engineers on educational visits to engineering works spending their time in photographing them. They appear to be paying more attention to 'getting a good picture' than noting structural details and listening to the explanations of the guide. They learn little from the photographs which are most probably put away in a drawer and soon forgotten. An engineer responsible for supervising, for example, deep excavation works must learn to look for tell-tale signs which may indicate overstressing, such as shearing of bolts in concrete segments or bowing of a waling in a sheeted excavation. The appearance of local seepage on the inside of a sheet pile wall may be the result of a split clutch which could have disastrous consequences.

The young engineer should take every opportunity that comes his way to practise close observations of such details. This is the best way to develop the necessary experience and confidence to take appropriate action when things go wrong or to institute beneficial cost-saving measures.

At the time I was visiting MacAlpine's work I did not appreciate that the formwork for the massive substructures had been designed at their head office and that the young site engineers whose work I so much admired were simply 'line and level' men, mere dogsbodies at the beck and call of the sub-agents and foremen. Later I appreciated that working with a medium-size contractor could give a better experience and higher responsibilities for the junior staff.

Although attracted by the outdoor life in contracting, I realized that a competent engineer must be experienced in design as well as construction. During the 1930s several of the larger county councils were undertaking the design and construction by direct labour of large highway and bridge projects using up-to-date heavy construction equipment. This seemed to me an ideal way of combining design and construction and in 1938 I obtained an appointment with the main drainage department of the Middlesex County Council. However all career ambitions were put on hold in the following year when the long-expected war came and I was seconded in 1940 to work on RAF and USAAF airfields in the UK and later in the Middle East. Before this I did have the opportunity in 1939–40 to put into practice the knowledge gained in studying MacAlpine's work by constructing tunnelled air-raid shelters and other blast protection work using direct labour.

War-time work on the airfields consisted mainly of laying miles of concrete runways and taxi-tracks. Although it could be classed as heavy construction it was technically undemanding; however problems subsequently occurred with some pavement failures under operating conditions with heavy bomber aircraft. I became acquainted with the then newly-developed science of soil mechanics and was able to put it into practical use in understanding the problems.

This led, after my return from the Middle East, to an appointment in the soil mechanics section of the Wimpey Central Laboratory. By the end of the war Wimpey had grown to become the largest building and civil engineering contractor in the UK. Their laboratory had been established in the 1930s for the testing and development of asphaltic road paving materials. In 1943 Dr Leonard Murdock was recruited from the Building Research Station to expand the organization and to undertake the development of non-traditional building methods with the aim of overcoming the expected problems of post-war shortages of skilled labour and the traditional materials of construction. Collaboration with the structural design department of the company led to the development of the no-fines concrete system of building construction, which was used successfully on a large scale for local authority housing throughout the UK. A system of low density in-situ concrete construction was developed for tropical housing and used in Kuwait, Iraq, Sumatra and Borneo.

Until its disbandment in the late 1980s the Wimpey Laboratory had expanded to cover a wide range of services including soil and geological investigations, testing of concrete, asphaltic and building materials, hydrographic surveying and the construction and testing of hydraulic models and large-scale structural assemblies. Employing some 400 engineers, scientists, technicians and drilling operatives, and providing services both to the parent company and outside clients, the organization was unique in the building and civil engineering industry.

My first appointment with the laboratory was as a foundation engineer; I later became technical manager and finally a director of Wimpey Laboratories Ltd. At the time I joined in 1947 I saw the appointment only as a stepping stone to achieve my long-cherished ambition of becoming completely involved in heavy construction, but this ambition was not fulfilled in the way I had intended. By this time I was married with a young family, and Dr MacGregor of Wimpey persuaded me of the advantages of following a settled career in the laboratory.

The work was fascinating and in many ways fulfilling as I became closely involved with many of the company's design and construction projects in the UK and around the world. I also came into contact with some of the eminent consulting engineers whose reputation had been made in the pre-war years.

During a discussion at the Civils on a paper on one of the early nuclear power stations, Lord Reith (who had qualified as a civil engineer) said that he would rather have been in charge of constructing such works than to have become the director-general of the BBC. This reflects my own feelings. The only material evidence of 60 years as a civil engineer has been the publication of two books. To paraphrase George Bernard Shaw's well-known saying, 'He who can does, he who can't writes a book.'

In *Middlemarch* George Eliot wrote of the engineer Caleb Garth who

> ...*often shook his head on the value, the indispensable might, of that myriad-faced labour by which the social body is fed, clothed and housed. It had laid hold of his imagination in boyhood. The echoes of the hammer where roof or keel were a-making, the signal shouts of the workmen, the roar of the furnace, the thunder and plash of the engine, were a sublime music to him; the felling and lading of timber and the huge trunk vibrating star-like in the distance along the highway, the crane at work on the wharf, the piled-up produce in warehouses, the precision and variety of muscular effort wherever work had to be turned out – all these sights and sounds of his youth had acted on him as poetry without the aid of poets, had made a philosophy for him without the aid of philosophers, a religion without the aid of theology.*

It seems to me that this passage epitomizes, in words far better than I could write, the satisfaction which I have found in my chosen career, and explains why the romance of civil engineering, particularly of heavy construction, has always been the dominant influence rather than its scientific interest.

Michael John Tomlinson

FICE, FIStrucE

Born in 1916 in Scunthorpe, North Lincolnshire, and educated at the local grammar school.

After 29 years with the Wimpey Central Laboratory he retired in 1976 to set up his own practice as a consultant specializing in foundations and earthworks. He has been involved as an advisor in a number of major projects including construction facilities for North Sea oil platforms; the River Lagan weir in Belfast; the viaduct sections of the Jeddah urban ring road and the Jeddah-Mecca expressway; and the Jamuna River Bridge in Bangladesh.

He is the author of *Foundation Design and Construction* and *Pile Design and Construction Practice*.

How engineers engineer

Sir Alan Harris

Problem solving?*

At a recent joint meeting between the Royal Engineers and the Institution of Civil Engineers, extensive consideration was given to the training of engineers, consideration which was explicitly intended to cover both service and civilian engineering – indeed, admirable accounts of actual experience of training were given from both sides. Lists were available of the matters to which RE training aimed – fully detailed, and reminiscent, somewhat, of aiming at the end of a barn with a bowl of rice – good cover of the target, but little penetration. It might be worthwhile endeavouring to put some near-philosophical order into what engineers are trying to do, what is needed for them to do it and how they might set about doing it.

To start off with, they are not problem-solving. They meet problems along the way and sometimes they need to solve them; at other times they change the problem so that it can more easily be solved or, better still, they dodge it.

The place of problem solving may, perhaps, be illuminated by considering two popular recreational problems, the chess puzzle and the crossword. Both are connected with activities of the practical intelligence, the first with the game of chess and the second with the use of words. Both are useful exercises. The chess puzzle exercises the mind in the use of two-dimensional space by pieces with limited mobility, and the crossword exercises it in the meaning and spelling of words; I have known French students of English who sought to extend and deepen their knowledge of its words by grappling with crossword puzzles. Note, however, that these are still puzzles, problems created by one intelligence for solution by another; in both, familiarity with the mind of the author helps provide a speedy solution. More, they are not only subsidiary, but disparate to their basic activities, which are the waging of competitive games on the chessboard and the use of the English language. Both these activities are imaginative exercises of the intelli-

* This article first appeared in the Royal Engineer's Journal, vol. 110 no. 2.

gence; in the puzzles, far from the creative imagination being called upon, they deal with a rigidly-defined construct from which the mind has to work back.

Engineering operations

Let us sketch out the sequence of intellectual processes involved in a typical engineering operation. Problem solving will be seen to take its subsidiary place.

Such an operation is a function of the practical intellect, devoted to some useful end (as distinct from the speculative intellect, devoted to the search for knowledge for its own sake). That useful end is typically of two sorts, making a useful thing or contriving action to achieve a useful outcome. Both form part of engineering; the second is, perhaps, more peculiarly part of military engineering, in that action is essentially linked with time. Both are subject to a preliminary activity called variously 'design' and 'planning'; the words are virtually synonymous, and imply asking such questions as 'what is to be made or done, how and what is needed for the purpose?', and terminate with instructions adequate for execution. The sequence of processes entailed in the procedure of carrying out engineering works or operations correspond, though sketchily, to the time-honoured formula for the 'appreciation of a situation' which went something along these lines: 'purpose' (nowadays, doubtless in deference to mid-Atlantic English, it is mission – we shall see later that the difference between a purpose and a mission is substantial), 'factors, courses open, courses adopted, plan.'

Let us set against this the stages typical of an engineering operation, whether of planning, or of design.

Appreciation of the task

This comes first. It is rare that an engineer is master of what the task is to be; he is usually acting under instructions. Nonetheless, his role begins now; his client/ superior officer will have an idea of what he wants, but the engineer may know more of what he could have; it is the engineer's responsibility to explore what is really needed, and he will not hesitate to return to the originator if new possibilities are revealed later in the process. It is here that the purpose/mission distinction becomes clear. A purpose leaves the engineer free to contribute; a mission is an order which the engineer has to execute by such means as he can. The mission is indeed a problem which the engineer must solve; a purpose leaves him free to make an engineering contribution to deciding what is to be done. It is good that engineers start by rebutting the quip that they know everything about what they are doing except 'why'. Their starting point should be a clear understanding, not only of what they are to do, but why.

Information

He must gather all the information he can about the task, starting with the physical data, whether geographical, meteorological or tidal, then passing to the availability of materials, labour and plant and including all matters concerned in the total function: what are the likely unspecified loads, what is the probable ill-use or misuse, are changes of use to be expected?

Conception and appraisal of possible schemes

Now the real work starts. The engineer meditates on all the knowledge he has gained, seeking in his mind a concept of a scheme which will satisfy the requirements of its intended function. Sometimes an idea will spring to mind of startling completeness; sometimes it doesn't. So it goes. The computer can help greatly in visualizing ideas and in facilitating their development. But without that intense meditation nothing will happen.

Usually, there will follow a string of ideas which must be subject to critical appraisal. Good engineers are as notable for the acuity with which they examine their own ideas as for the fecundity with which they produce them; both creative and critical faculties are equally necessary.

Having conceived an idea, the engineer will appraise it against three criteria: functional soundness, economy in execution and reliability in use. Questions of functional soundness may well prompt a return to the promoter, either negatively by indicating unrealistic requirements, or positively, indicating unexpected possibilities. As for economy, the designer must know how the work is to be built; in general, if he can think of one way, there will be six; if he can think of none, then none there probably is. But economy is more closely related to simplicity of execution than it is to quantities of material. A method of construction that proceeds simply with successive operations following one another smoothly with no backtracking or interference is the best assurance of economy.

At this stage there will be few detailed costings or structural calculations; the engineer will rely on his judgement and experience. It is essential that, having fixed in broad terms a method of execution, there are no details which are inconsistent with it. Detailed analysis of some minor features may be necessary to ensure that they are indeed so consistent.

This process is repetitious; the time will come when a decision must be made. This requires judgement, not only as to its matter, but also as to its timing: the decision can be rushed with dire consequences to the design, or it can be delayed too long and miss a deadline. Decision is a matter of will, of character. Some panic and decide too soon on a project of huge and unresolved complication, when a little more time would enable the scheme to be torn up and started again, with better results. Others are unable to resist the search for perfection, forgetting that one adequate scheme on time is worth many masterpieces too late. Moreover, working too much on a scheme can over-elaborate it; there is a right time to decide. It is, like the elephant, easier to recognize than to define. Experience helps.

But the moment of decision is the capital moment; hitherto it has been all in the mind, but now harsh reality threatens. Now the ponderous wheels are set in

27

motion which will end in construction. There is much work yet to be done in the development of the scheme, work which can go wrong and wreck a right decision, but nothing now can save a wrong one. The significant decisions concerning structural form, material and method of construction, have now been taken.

Preparation of the instructions needed for execution

This is what remains. (In what follows, consideration has been restricted to the design of a structure; while the planning of an operation is analogous, it defeats clarity to seek words which apply to both construction and action.) Such instructions may require detailed checking of the strength of structural elements – a process often mendaciously referred to as 'design' – it is no such thing. Design is not working out the dimensions of a beam, the significant design decision was to have a beam. But while the competent engineer will have an approximate idea of the size of the elements of his structure, it is now necessary to bring to bear the full armoury of structural analysis to foresee how the structure will behave in practice and to enable it to be so detailed that it will function as required.

However precise the figures, they can never serve as more than an aid to judgement. Accident-prone structures can be recognized: the detail which is adequate if perfectly executed, but which is catastrophic if inexactly done; the possibility of progressive failure. The achievement of safety is an art which is nourished by knowledge of the site and its processes.

It is a rare structure to which someone does not, at one stage or another, entrust his life; the extreme example is the footbridge over a chasm. The avoidance of loss of life due to structural collapse is, indeed, a major factor. Were it an absolute requirement, however, all things built would approach the condition of a pyramid, be hugely expensive, take an eon to build, be utterly stable and no use to anybody. In practice, total safety in this sense is unattainable, spend what one will on design, analysis, selection of material, supervision and testing of construction, application of no matter how many shelf-loads of codes and regulations. It is not enough, moreover, that the work should not collapse. The idea of safety is best subsumed in the more general concept of reliability, whereby the work can be relied upon to continue to function as intended for as long as is required, a reliance which includes, needless to say, the avoidance of fatal collapse.

Structural analysis requires three operations:

1. **The definition of an action on the structure**.
   Such an action may be a load, the effect of environment such as variation in temperature or humidity, the wear and tear of use and likely misuse. It is very rare that such actions are capable of precise definition. The only well-known instance is an open-topped water-filled reservoir which can be overloaded only by the unlikely replacement of water by a liquid of higher density. Otherwise, loads vary over a very wide spectrum with a frequency which typically follows the familiar Gaussian bell-shaped curve; it is extremely difficult to determine a

load which will never be exceeded. In consequence, the prudent engineer will cast a sardonic eye over load values fixed by official regulation; they may or may not be satisfactory from the point of view of legal defence, but no prudent engineer will assume that they can never be exceeded. With some loads, of course, such as those caused by wind and wave, the variability is well-known and every effort is made to assess probabilities by statistical analysis and to design to provide strength on a basis of probability. In addition, loads in practice are of three sorts – steady and of long duration, when rupture load is the criterion; repeated when elastic stress leading to fatigue failure must be guarded against; and dynamic load, when energy absorption and natural frequencies of vibration count. Nor should change of use be lost sight of: when an office is turned into a library, floor loadings are increased. Perhaps the engineer should see to it that the load for which the floor was designed should not be forgotten but enshrined in some official documents.

One further predicament with which the engineer has to wrestle should be mentioned, and that is the coincidence of extreme loads. The most familiar is the concurrence of a high tide with a storm-force wind from the most unfavourable quarter, but there are others.

2. The analysis of the effect of the action on the structure.

The means of structural analysis are immensely powerful these days, but the engineer should not be misled by the seeming accuracy of eight significant figures emanating from a computer. He should remember that the answers are no more accurate than the questions, i.e. than the loads on which the calculations are based. Moreover, he will get answers only to questions which he asks. If he has never heard of buckling, the computer will not tell him. As a general rule, the operations of a computer are of extreme accuracy within their terms of reference, but if they are inexact, they can be wildly inexact. The prudent engineer will thus generally accept only such answers as appear to him to be about right.

Moreover, the huge prestige of mathematics was gained in the service of disciplines whose needs are not ours. Though non-linear functions are beginning to be susceptible to mathematical analysis, most such analysis is based on the assumption of perfect elasticity, i.e. that stress is proportional to strain.

3. The comparison of the effect of the action with a criterion of adequacy.

This is just as important as the evaluation of the action, and is just as often neglected. There are many criteria in use: safe stress, elastic stability, and ultimate collapse are chief among them, but they must always be viewed in the light of the designer's knowledge of the nature of the load. (See paragraph 1 above.) Until very recently, triaxial stress conditions in large concrete masses could, indeed, be analysed, but would then be looked at only with blank curiosity because there was no available distinction between combinations of stress which were safe and those which risked rupture. This inadequacy has now been made good, but it is necessary to bear in mind the total lack of meaning in figures which appear at the end of a series of mathematical computations, in the absence of a criterion by which to judge them.

In parallel with these processes, there continues the preparation of the detailed drawings of what is to be built. The purpose of these drawings is to give instructions to the builders, and it is good to know what instructions are necessary and how they are best expressed for ready comprehension. This presumes a knowledge of construction methods.

The specifications and contract documents are part of the design and should complement by description and definition the information contained on the drawings.

The designer's work has not finished when a roll of drawings and a bundle of documents have been handed over. It is a rare job which is free of alarming incident, and the designer may well be alone in the ability to assess its importance and to deal adequately with the problems it raises.

Classic military 'appreciation of a situation' compared with engineering design

---

Purpose/mission	**Appreciation of the task** Keen enquiry as to all aspects of the task and the need to be satisfied (including why), leading, perhaps, to advice as to what is available beyond the knowledge of the promoter.
Factors	**Collection of relevant facts** All possible information relevant to the task and to the total function is assembled, together with information relating to means available for achieving result.
Courses open	**Conception and appraisal of possible schemes** General ideas of a project will be imagined which will then be appraised in terms of function, economy and reliability in use.
Course adopted	**Decision** The crux of the operation.
Plan	**Instructions for execution** The examination of the project in detail sufficient for instructions for execution to be drawn up, which implies the thorough checking of the project for soundness, strength and reliability. **Collaboration during execution** The author of the design must be ready to advise during execution.

Striving for excellence

The engineer is expected to produce a work which fulfils its function satisfactorily, at a price and a time within the given target; he will feel obliged by the very magnitude of his task to seek not merely adequacy but excellence in the final product. Excellence is to be recognized by unity: the work is seen to be one thing, with its parts visibly serving the whole, not just a heap; by simplicity, that air of ease which is typical of all good art; by necessity, perhaps the secret of excellence. It is paradoxical that a work in which there is nothing that is not seen to be necessary produces an effect, not of privation, but of magnificence.

These characteristics are not to be consciously striven for, but when the designer has striven for goodness, has done his sums right, has struck out all that frippery which looked so pretty, when he has done all this and looks at his work and sees that it is single, simple and spare, he knows he has attained excellence. He is God on the seventh day.

The process of design is eminently a matter of management and teamwork, remembering that good design is not the result of consensus, but of command. A design produced by a committee is proverbially bizarre. There should be discussion by all means, but dispute is to be avoided – it leads to emotional commitment and confusion.

How is the engineer to be equipped for this task? The requirements are formidable. He must first (and this is basic) know his materials: what they are made of, how they are made, how they are shaped, formed, assembled and erected; how they behave under load and under all the various agencies of ruin which begin to act as soon as construction is complete; and, finally, how, in the long run, they fail (grass will conquer in the end).

He must study and have a knowledge of great works of history. 'I pass for a man of quick wit,' said Napoleon, 'but it is rather that I have much reflected.' To this he must add his own firsthand experience, of which it has been said by Detoeuf, a famous French mechanical engineer, 'Real experience is secret. It is born of small, daily incidents; small, repeated errors; small, renewed successes which, by their number, mark laws to which habit submits. It is not thought out, it is lived; one lives with it unconsciously, as with gravity.'

'Time spent on reconnaissance is seldom wasted,' but on condition that one knows what one is looking for, recognizes it when seen, and is able to use it. Experience of works requires cogitation over observation to be of value.

So, we have study, experience and responsibility. (There is no replacement for necessity; it is surprising how fast one learns when responsibility and necessity drive.) A foundation of maths is probably essential, not so much for calculating as for understanding a language; there are ideas which can be expressed only with difficulty in other than mathematical terms.

This sounds all very burdensome, but there is an historical line of research open.

An antique guide book (c. 1923) refers to the Albert Hall and the contemporary Albert Memorial. The former, designed, it says, by Lieutenant Colonel Scott RE for a joint stock company, cost £200,000. We know of its difficulties; designed before the beginning of the science of acoustics (the Greeks had inklings of the art, at any rate for the human voice out of doors), leading to that famous echo which prompted a French musical friend to remark 'The only way for a British composer to have his composition heard twice in public is to have it played in the Albert Hall', but there it still is, echo removed and immensely successful.

Just in front, says the guide book, is the Albert Memorial designed by the famous Gothic revival architect, Sir Giles Gilbert Scott, and costing £120,000, and 'far from commanding universal admiration.'

It is not widely known that, apart from the Albert Hall, the Royal Engineers undertook a very large volume of civil engineering works in the nineteenth century. There was the Coles building in the Victoria and Albert Museum; and elsewhere there was a sequence of dockyard buildings, mostly covered ship-building slips from Sheerness round to Devonport. Indeed, the Corps became an effective civil engineering arm of HM government. The instruction received at Woolwich by regular RE officers was highly technical and, at a time when cast iron was followed by wrought iron and then by steel, put them in the forefront of knowledge of these materials and leading practitioners in their use. At a time when there is much discussion as to how engineers should be educated, some historical examination into how it was done at 'the Shop' might be worthwhile – it clearly worked.

In conclusion, one quotes Countess Morphy, the authority on traditional British cooking. 'Plain cooking,' she said, 'is not to be entrusted to plain cooks.' We have seen how engineering works can excel by their simplicity, such works are not produced by simpletons.

Sir Alan has been an inspiration to many in the profession. One of his most profound statements was 'let the subconscious take over'. In other words, worry away at a problem, then sleep on it. In the morning everything will often seem much more straightforward!

Colonel Emeritus Professor Sir Alan Harris

CBE, FREng, FCGI, HonDSc, BSc(Eng), FICE, FIStructE

Born in Plymouth in 1916, of a maritime family, the author joined the Corps in June, 1941. Volunteering as a parachutist, a damaged knee and medical downgrading saw him posted to Chatham in 1942. He volunteered for port construction and was posted to 933 Port Construction and Repair in 1943, where he undertook courses in diving and underwater bomb disposal. As OIC advance party in Normandy, he landed in Port en Bessin D+2; was detached as OIC divers, Mulberry B with fleet of French fishing boats; in August was in Ostende; and was OIC advance party Xanten am Rhine for construction of a timber-piled Bailey bridge built in cantilever. At the end of the war he worked with others to set up a control commission team to open up the Dortmund-Ems canal. He was mentioned in despatches and received the Croix de Guerre.

Amongst the author's many accomplishments since the war, is service as a colonel in the Engineer and Transport Staff Corps and being elected Professor (now Emeritus) of Concrete Structures at Imperial College. He was a founder member of consultants Harris & Sutherland and pioneered many prestressed concrete structures. For the latter he took as his inspiration the great Frenchman M. Eugene Freyssinet whom he knew well. He has been President of the Institution of Structural Engineers and a Vice-President of the Institution of Civil Engineers.

The chronology of a career in engineering

Peter Campbell

The early years 1943–52

In 1988 I delivered a Presidential address entitled *Structural engineers: some of their wider responsibilities*, and this essay provides me with an opportunity to describe how that approach has influenced the development of my career, which in a sense started in the middle of the Second World War.

My background was very ordinary and working-class, but that was influenced by the appearance of my stepfather, the late E.L. Campbell FRICS FIStructE, a professional man, who came into my life in the early 1940s and arranged for me to attend Purley Grammar School for Boys, where my education started in 1943. 'Pop' as I always called him spent his professional life in the London District Surveying service, and encouraged me to visit his office regularly and go with him to visit construction sites when there was something interesting to see. It was these experiences that first sparked my interest in the world of construction.

I was in my early teens at the time when I decided to embark on a career in structural engineering. If anybody is to be blamed for that decision, then Pop is the one to blame, for he was the first person to encourage me and convince me that I could perhaps follow in his footsteps, but like him qualifying the hard way.

My school showed very little interest in its pupils unless they were clearly Oxbridge material, and the idea that one went to university to read engineering was certainly not encouraged. So during my first year in the sixth form in 1948 I decided to apply for a place at the Brixton School of Building, having been told that it was unlikely that I would be successful, since at that time preference, quite rightly, was given to ex-servicemen. However, to my surprise, two weeks before the commencement of the academic year I received the offer of a place to read structural engineering for three years on a course that in those far off times was multi-disciplined, and I was to rub shoulders with architects, surveyors and builders, an aspect of my education that proved to be very important.

Training and career development 1952–58

My second piece of good fortune happened when I had successfully completed the course, and the late Charles Hockley, who had taught me at the school, arranged for me to be interviewed for a job with Ove Arup & Partners. My most important memory of that interview with one of the then partners, was when a chap in shirtsleeves marched into the large cupboard which passed for an interview room, looked over the shoulder of Andrew Young, to read his notes, and exclaimed, 'not another bloody Peter' and stomped off. I was very apprehensive! Subsequently I learned that it was the late Peter Dunican, who was to become not only a friend but an important mentor, after I joined the practice as staff member number 34 in 1951.

At the tender age of just 20, having enjoyed a very sheltered existence, I was suddenly surrounded by people from the colonies, Germany, Poland, Scandinavia and so on, many of whom had spent the war years defending people like me, who had then been educated at university, and by my standards were very mature and able colleagues. I grew up very quickly, and received the sort of further education and training of a calibre that I do not believe was available anywhere else; good fortune for which I have always been grateful. I was imbued with the Arup ethos and philosophy which has been the foundation on which the whole of my career has been built.

It soon became apparent that I needed to enhance my technical ability at a time when I met the late John de C. Henderson, who was teaching postgraduate students on the DIC course at Imperial College. One was required to have an honours degree in engineering if one wished to attend that course, but John nevertheless encouraged me to apply for a bursary and a place in 1953. On both counts I was successful, spending the 1953/4 academic year studying concrete technology and advanced structural analysis. This was an opportunity to which I applied myself and benefited in terms of the future direction of my career, not least by virtue of my attending Professor Skempton's lunch-time lectures on the history of civil engineering.

On my return to Arups, from whom I had been given leave of absence, I rejoined the group led by another important and much admired mentor, Andre Bartak, who shortly after became very ill and was inevitably due to be away for some time. While he was making a good recovery, I assumed responsibility for the work of the group. This was important since one had to apply skill not only to the day-to-day running of the numerous projects, but also the management of the people within the group, and its representation in the firm.

The birth of private practice 1958–69

Ian Reith had also studied at the School of Building, and then joined Arups one year ahead of me, and it was in the mid 1950s that we decided to set up our own practice in due course. In 1958 I started to do simple jobs in my spare time, which culminated in working with one of Arup's partners on the design of a small hospital in India for a Christian charity that had few resources. They were only able to pay a nominal fee for the work that we did, but which was appreciated. For me it was an important lesson about the need that exists to use one's abilities not only for personal betterment, but also as a means of assisting those less fortunate.

Many of the people I knew professionally encouraged me to start the practice I was now determined to have, and I left Arups in 1962 in order to take those first tentative steps. Most of the people who had encouraged me to take that step and who had promised their patronage failed to deliver, with the exception of an architectural practice called Lyons Israel & Ellis who soon gave me some work. I slowly started to make my own contacts, and work began to flow. But it was in the mid-1960s that there was an unexpected development.

One of Arup's principal patrons, since before I joined the firm, was the architectural practice of Maxwell Fry and Jane Drew. The story as it was told to me

FORESTDALE FOOTBRIDGE, ADDINGTON.
The Forestdale Housing Estate was at the time the largest private housing development undertaken, and some 30 years ago the developer had recognized the value of separating completely pedestrian and vehicular traffic. The footbridge carries the pedestrian route over the main vehicular traffic's entry to the estate, and comprised two in situ *concrete Piloti supporting a wholly precast concrete deck. This resulted in traffic entering the site being restricted for only three hours, and the construction cost was half what the developer had allowed for in the budget.* Designer: Peter Campbell.

was that when Arups announced the establishment of Arup Associates, a multi-professional practice including architecture, it produced a sharp reaction from Maxwell Fry, and as a direct result of that I received one fine afternoon a telephone call requesting my urgent appearance at the offices of Fry Drew. I had worked on their projects at Arups for ten years, and was consequently well-known. It was I believe the reason why I was summoned, and as a result of what proved to be a temporary schism in the Arup relationship I was asked to accept an appointment to carry out a series of projects in Mauritius, starting with the Government Centre which was required for independence celebrations in 1968, and a large hospital project. This appointment proved to be a watershed and provided a track record for overseas work which was to soon develop, resulting in the appointment in 1969 for a World Bank Education Project in Trinidad and Tobago with Architects Co-partnership. This project involved the design and construction of 22 comprehensive schools for 1000 pupils each and a teacher training college.

THE GOVERNMENT CENTRE, PORT LOUIS, MAURITIUS, PHASE 2.
*Sir Seewoosagar Ramgoolam the Prime Minister commissioned Messrs Fry Drew Knight &
Creamer, architects, and my practice, to design numerous projects on the island of Mauritius.
Phase 1 of the Government Centre was constructed in 1968, for their independence
celebrations. Concrete is the primary structural material, but the cement and steel is all
imported, aggregate being obtained by crushing the basalt rock. Phase 2 was designed specifying
carefully controlled board marked concrete. The island is frequently visited by cyclones, and as a
result the buildings were designed to cope with wind velocities of 160 mph, and gust velocities
of 180 mph. These requirements impose strict conditions on one's approach to structural
architecture in that environment.* Architect: Messrs Fry Drew Knight & Creamer.

The partnership develops, 1969–76

Ian Reith, having spent some time working with other consultants and contractors, joined me in 1969 as my partner, and in so doing, made the undertaking that we had entered into a decade earlier a reality. 1969 was clearly a busy year, for it was the year that we decided to participate in a firm that we established with architects, Geoffrey Salmon & Associates, and quantity surveyors Nigel Rose & Partners, with initially an office in Dubai which had a resident partner. Between 1969 and 1978, we did many small and medium-sized projects in the Emirates, but when the bubble eventually burst, Campbell Reith & Partners pulled out.

During this time, in the early 1970s, several experienced engineers applied for posts with the practice. Notable amongst these were Bill Hill, Stuart Goodchild, Mark Kaminski and James Tasker, all of whom have been responsible, with other partners who were appointed, for the development of the practice which is for me a source of great personal pride. Since the practice was no longer one man deep, and all of my partners were establishing their own relationships and client base, I decided to seek their authority to spread my wings and involve myself in other but kindred activities.

If one has a modicum of success in one's career, there comes a time when it is appropriate to share your experience with others if the net effect is to bring about benefits to your local community, society at large, your profession and not least, enable you to develop an awareness of how one can be of service, and the communication skills that are essential if help, advice and support are to be provided. Many people who know me today will not be aware of the difficulties I had in communicating, due to shyness and a lack of confidence, which had to be worked at, and were an impediment in my early career. I suspect this is not uncommon in many of us.

The introduction of wider interests 1976–7

My involvement in extramural activities started when I was appointed a Justice of the Peace in 1976, having first obtained the approval of my partners. Since then I have on average sat in court about 40 times a year. This work is important rather than enjoyable and, I believe, is work that engineers by virtue of their training are well-equipped to do. Sifting and assessing facts and information is central to problem solving, and as a magistrate the problem is to establish either the guilt or innocence of the defendant before you. Sometimes it is even possible, rather than to inflict punishment, to take the opportunity to give people the guidance that they clearly need.

I soon directed my attention to my profession, that of a structural engineer. In 1979 I was proposed as a prospective member of the Institution of Structural Engineers Council, and was duly elected. I recall with pleasure receiving a telephone call from past president Peter Dunican immediately the result of the vote

THE MAURITIUS HOSPITAL.
This is located in the centre of the island, and was by UK standards a district general hospital which was to ultimately provide 750 inpatient beds, with maternity and A & E facilities. The site is surrounded by sugar cane plantations, with rugged mountain scenery behind. With high velocity wind forces, one of the factors that influence the design is the need to protect the building from flying debris. Because of the brightness of the light we strictly restricted the size of windows and protected them where possible with roof overhangs to limit the perforation of the external envelope. Architect: Messrs Fry Drew Knight & Creamer.

was announced, congratulating me on coming top in 'the Council examination!' Peter was a generous chap, having not only given both Ian Reith and me encouragement, but passed on several clients to us both in the early days. My work for the Institution was assisted and encouraged by Institution secretary, the late Cyril Morgan, and in due course I found myself faced with the real prospect of becoming president. I was installed in that office in October 1988 and served for the 1988/9 session. During the previous nine years I was involved in numerous standing committees, task groups and working parties, such that on assuming office I was well-versed in the Institution's activities and affairs.

It is difficult to assess the effect if any, that your period in office had either on the Institution or its members. I did however take a sabbatical from my practice and devoted my full attention to the work of the Institution. It was an enjoyable and worthwhile experience, but others will have views as to the value of my year in office.

STAGHILL COURT, UNIVERSITY OF SURREY, STUDENT RESIDENCES.
Staghill is part of the university campus, and is the long slope below the cathedral, which was an unstable clay slope of 1 in 7. Much of the first phase of the university had been completed when I was commissioned to design residential accommodation on this site for 440 students. All of the existing buildings were supported on piled foundations, a solution that I decided to review.

In simple terms the stability of clay slopes is a function of the slope, the moisture content of the clay, and the extent to which additional load is applied to the slope. The structure/soil interactive solution that I proposed for this site of approximately 2.5 acres was to first regrade the slope to 1 in 9, by removing an overburden equal to the load that would result from the construction of the buildings; and then to surround the site with a 5.0 m deep granular filled moat fed with shallower land drains parallel to the slope, which all drained into the newly-formed artificial lake. The whole area was then covered with a stepped raft on the surface 250 mm thick RC, with 4.5 m treads and 600 mm risers on which was constructed low rise buildings arranged like a traditional Italian hill village with passages which opened on to small courtyards. Holes were left in the raft to receive planting. The raft was concrete as was the paving, and the buildings were made entirely using Forticrete concrete blocks for vertical support.
Architect: Messrs Maguire & Murray.

STUDENTS' UNION BUILDING, UNIVERSITY OF SURREY.
This building followed Staghill Court, and is situated at the bottom of the site below the residences. Raft foundations were again used on this building, but because of the flatness of the site much simpler drainage was necessary. Architect: Messrs Maguire & Murray.

The Museum of Concrete and the ACE Agreement 1984, 1979–89

Before moving on I want to mention another project that started in 1979/80. At that time I was involved with the Concrete Society, and as I have said my interest in engineering history had been sparked by the lectures of Professor Skempton, and on making enquiries, I found that nowhere was the history of concrete covered in our museums. I resolved to remedy that omission.

I have never understood why the Society and the cement and concrete industry were unprepared to assist me with that endeavour, and as a result, that resistance made me even more determined to press on with my plan. With the support of Amberley Museum and a small group of like-minded friends, in particular Christopher Stanley, I refurbished a building that the museum made available to me and then collected all the material which enabled me to proceed with the setting up of the History of Concrete exhibition, which was formally opened by Sir Ove Arup in June 1981 and continues to be an important source of information about the historic development of the use of concrete.

Acknowledging my limitations, I have always resisted taking on more than one important task at a time, on the basis that if you are asked to take on certain responsibilities, you should expect to give those responsibilities your undivided attention. During my presidential year I was approached suggesting that I might consider being put forward for the Council of the Association of Consulting Engineers. This was somewhat of a surprise, since I had successfully ruffled a few feathers in the ACE, when in 1983 I set up a working party to draft Agreement 3 (1984), an agreement, first published by the ACE as a pilot edition, that we believed was an improvement on the 1981 version drafted by others. This belief, in a short period of time, was confirmed by the use of the document by the membership.

The ACE, RedR and the Construction Disputes Resolution Group, 1989–92

Having completed my term of office at the Structurals, I was elected for a period of three years as a member of the Council of the ACE in 1989, at a time when proposals to restructure the ACE were being resisted. In 1990 I was invited to take the role of vice-chairman of the ACE, which I was honoured to accept, then to be told that my task for the year was to have another try at persuading the membership of the advantages of membership by firms rather than by individuals. In 1991 I was installed as chairman of the ACE, but the work of securing the members' approval of the proposed reorganization was ongoing, and largely due to the efforts and support of Hugh Woodrow, the ACE's chief executive, the changes were approved during my term in office.

As vice-chairman I was given the opportunity to launch the Patrons Scheme on behalf of RedR, Engineers for Disaster Relief, at the annual dinner of the ACE in 1991. Shortly after the successful launch of that scheme I was invited to become a vice-president of RedR, and continue to do all that I can to raise funds and promote the valuable work that RedR and all of its volunteers do. Engineers generally tend to whinge about their lack of status in society, but I believe that you earn status, and status is accorded on the basis of what you do and what you achieve. RedR is a fine example of the humanitarian work carried out by dedicated young engineers, supported by their firms, in areas where disaster strikes, often in considerable personal danger. It is activities like these that earn respect and understanding about the quality of my profession and the people of which it is comprised.

RedR International was established in Switzerland in 1995/96, to be close to the international agencies, and to act in an advisory capacity when other countries decided to set up their own RedR groups. New Zealand, Australia, USA, Canada, Nepal, and others, have or are in the process of achieving that goal.

At the time of my RedR involvement, Kenneth Severn, a distinguished colleague and friend, suggested that we establish a branch of a group that had re-

ST AUGUSTINE'S CHURCH, TUNBRIDGE WELLS.
Maguire & Murray led the movement to change the manner in which church worship was conducted, from the traditional axial arrangement to one which is circular, and which brings the congregation close to and surrounding the sanctuary. This church is one of many that were designed to achieve this objective, but all were very different. However, steel and timber were the predominant structural materials that were used in these projects, to provide large unobstructed spaces. Architects: Messrs Maguire & Murray.

cently been set up in Canada by a colleague who is a contractor, called the Construction Disputes Resolution Group (CDRG). We were both concerned about the state of affairs that surrounds dispute resolution in the construction industry, so we resolved to promote the amicable dispute resolution process which has its roots in mediation.

The driving force has always been Kenneth Severn, but he asked me to act as chairman of the group, whose main achievement has been the promotion of these ideas by writing and the organization of seminars designed to spread the word. We have a list of practitioners who are supporters of our approach, representing all factions of the construction industry, and who can be nominated when we are approached by parties who wish to appoint a facilitator who can assist them to resolve their dispute.

However, as a result of Sir Michael Latham's recommendation in his report entitled *Constructing the Team*, adjudication has been adopted by parliament as the preferred process. Nevertheless there is still scope for the amicable process to be employed either side of the adjudication, that became obligatory in 1998. This is to be encouraged, since if one of the parties will not accept the adjudicator's award, the dispute will have to be referred to arbitration or litigation, the position CDRG has campaigned to avoid!

Retirement, FIDIC and the Concrete Structures Group 1992/93

When a new business or professional office is launched, it is important to establish clear terms and conditions for all those that will be involved. Campbell Reith Hill embodied their terms and conditions in a partnership agreement. The agreement was drafted on the basis that the practice would be ongoing when the original founders, for whatever reason, no longer participated in the affairs of the practice. To that end it was therefore necessary to select future partners not only because of their quality, but because they fitted into an age profile that provided

THE HEADQUARTERS OF SLOUGH ESTATES, SLOUGH.
This building was constructed below the Heathrow flightpaths, and the client wished to be spared from intrusive aircraft noise. The construction is entirely concrete, in situ reinforced concrete for the primary structure, and precast concrete for the external cladding. One of the perennial problems of cladding concrete frames with precast concrete cladding is the question of compatible tolerances. In this project the tolerances applied both to the frame and the cladding, and were plus 0 minus 3.0 mm. All of the units fitted perfectly. It was however necessary to enliven the internal environment by the introduction of some electronic background noise, having effectively dealt with the external noise. Architect: Salmon Speed Associates.

THE ARUNDEL WILDFOWL TRUST.
The site for this building is located in the wetlands adjacent to the river Arun. The investigation of the site revealed that there was a firm crust some 3.0 m in depth overlaying 60.0 m of mud. Since Sir Peter Scott's budget could not accommodate the cost of piling, it was decided to float the building on a 2.0 m deep cellular concrete hull. More recently the building has been completed, the extension having been provided for in the original design. Sadly the beautiful backdrop of trees was completely destroyed in the 1986 storm. Architect: Mr Neil Holland.

for future promotion as senior colleagues retired. Retirement can occur between the ages of 55 years and 65 years, providing flexibility and a degree of personal choice. In 1992 when I reached the age of 60 years I felt that I wanted my younger partners to be in control of their own destinies, as I had been for a long time. It was therefore agreed that I would retire, and hand over the reins of the senior partner to Bill Hill, who is at least a decade younger than me. He has been a great success in his stewardship of the practice through a very difficult period in the construction industry.

1993 was another eventful year. FIDIC, the International Federation of Consulting Engineering Associations, required a representative from the UK to sit on the Executive Committee. At that time the ACE was the second largest member association, and by tradition had a place on the executive. As there were no other volunteers to take on that responsibility, I said that I was prepared to, and in 1993

45

THE BEDSER STAND AND KEN BARRINGTON CRICKET SCHOOL, FOSTERS OVAL.
The grandstand and hospitality boxes were constructed above the large sports hall that constitutes the cricket school, which is below ground. The substructure was constructed using reinforced concrete and piles, and the superstructure using steel and concrete. The site investigation predicted that as a result of the removal of the material to construct the basement, there would be an elastic vertical expansion of the underlying clay of up to 75 mm which had to be accommodated. To achieve this the basement slab was cast on 100 mm of Clayboard, cellular panels made from paper. It was discovered elsewhere when the construction was complete that in certain conditions of temperature and moisture Clayboard is subject to anaerobic degeneration which produces methane gas. This was the case on this project, and one of my partners devised an ingenious method of removing the remains of the Clayboard from beneath the basement slab (which is too detailed to describe here) to avoid the future risk associated with the existence of methane gas in a building. Architect: Messrs Darbourne & Dark.

was formally elected to serve for four years. During my term I was the link between the EC and the client consultant relations committee, and had special responsibilities for approving applications to be included on the registers of mediators, adjudicators and arbitrators, having particular interest in dispute resolution procedures. I was also involved in the provision of a new policy statement on the FIDIC approach to grand corruption, which resulted in the expansion of the FIDIC code of ethics to embrace the requirements of the new policy. My duties culminated in my involvement as a member of the organising committee for the 1997 FIDIC general assembly meeting. It was held in September of that year in Edinburgh and attended by 650 people from around the world. HRH The Princess Royal attended the opening session as our principal guest, and as president of RedR. Her attendance resulted in the raising of a substantial sum for RedR International.

In May of that year, I received a totally unexpected invitation out of the blue, to establish a trade association to represent the interest of a group that had no proper representation, namely the specialist subcontractors who build concrete framed structures. We called it Construct, an acronym for the Concrete Structures Group. I served as the founder chairman for three years, at which time I handed over the duties that had been assigned, to Stef Stefanou, the chairman of the John Doyle Group. I was asked to accept the ongoing title of President of Construct, and continue to be a member of Council in that role.

In the early days, I was anxious that Construct should take on a limited number of projects, and carry them out comprehensively and with authority. The first was to draft a manual on contractor detailing of reinforced concrete, that was designed to provide the specialist with the opportunity to make an input into the design process, and then detail the structure to facilitate the method of construction the contractor intended to employ.

Cardington and the Paviors Musuem 1993–97

The next project was very ambitious. With the assistance of a small working party we decided to promote the need for a major research project, to determine best practice in the quest for greater efficiency, quality and minimum cost in the construction of concrete frames. To cut a long and arduous story short, this resulted in the creation of the project known as the European Concrete Building Project (ECBP), an industry-led project, in partnership with BRE.

The next few years will see the Implementation Committee under my chairmanship organize the construction of up to four building frames at the BRE large building test facility (LBTF), at Cardington. The members of Construct, suppliers of cement, concrete, formwork, reinforcement and plant, the BRE and DETR, guaranteed that phase 1 of this project would commence with the erection of a seven-storey, four bays by three bays *in situ* reinforced concrete frame in January 1998.

The ECBP has been described as the largest, most comprehensive study of the design, construction and the whole building performance of reinforced concrete frames ever undertaken. It is planned to follow the construction of the *in situ* frame with a smaller innovative frame using precast elements glued together with ultra high-strength fibre reinforced concrete, and then a hybrid using precast and *in situ* concrete. Finally a wholly precast concrete frame will be constructed but that is in the future.

A full programme of process and performance investigations will be carried out and the results will be instantly available on a project-specific website which will be constantly updated. In addition regular bulletins will be published by BRE.

I have always been open about my view that the structures we design in the UK are generally over-designed and over-specified. This project is precisely the vehicle for proving whether or not such a view is justified.

THE BRE LARGE BUILDING TEST FACILITY, CARDINGTON.
The Cardington facilty under construction. It was finished in May, 1998.

I referred earlier to the History of Concrete exhibition at Amberley established in 1981. More recently, I was appointed a trustee of that museum at a time when the master of my livery company proposed the establishment of a museum of roads and road making. Since there was already a small example of early road building at Amberley, I proposed that the Paviors Museum should be situated there. Plans were drawn up by the project working party, although it was destined to take some years before the project was to be completed.

A large existing building on the site was extended, and a fine exhibition assembled, that details the hisstorical development of roads and road making in Great Britain, and contains a collection of plant, equipment and road furniture, together with models of well-known road schemes and road bridges. This department of the Amberley museum is called the Paviors Museum of Roads and Road Making.

Writing as a medium for communication

During my career I have written many articles, papers and reports. In 1982 I was asked to contribute a chapter in the book entitled *Structural Engineering – 200 Years of British Achievement* that was written to celebrate the 75th anniversary of the Institution of Structural Engineers. The commemoration of the centenary of the birth of Sir Ove Arup occurred in 1995 and I was asked to write an essay for a small book entitled *Ove Arup 1895 to 1995* that was associated with the Ove Arup exhibition mounted at the ICE.

In November 1997 the publisher of this book produced a book called *Construction Disputes: Avoidance and Resolution* that I had the privilege of editing. It is, I believe, the first book to be published, written by construction professionals for construction professionals, which describes all of the various methods of dealing with disputes when they arise, or appear on the horizon. The book was also intended to provide a reference book for use in education, and to that end it has been successful.

Finally, this contribution is intended to describe my career, particularly to young people who are starting out on theirs, to illustrate how satisfying is the work of a structural/civil engineer, and the extent to which one can achieve personal fulfilment from your professional activities in addition to serving your profession, your community and society at large.

Eur Ing Peter Campbell JP

FREng, FCGI, DIC, FIStructE, FICE, FIMarE, FIHT, FIDE, FRSA, Hon MCONSE

Educated at Purley Grammar School, the Brixton School of Building and Imperial College of Science Technology and Medicine.

Joined Ove Arup & Partners in 1951–1962, then started the consultancy practice Campbell Reith Hill. Senior Partner (1962–92), now Consultant. Justice of the Peace. President of the Institution of Structural Engineers (1988–89); Chairman of the Association of Consulting Engineers (1991–92); Chairman of the European Consultants Network MERGE (1990–); Chairman of the Construction Disputes Resolution Group (1990–); Chairman now President Construct, the Concrete Structures Group (1993–); Founder and Chairman of the Implementation Group for the European Concrete Building Project at BRE Cardington. Chairman of the drafting committee for the National Concrete Frame Specification. Civil Engineering Assessor for the Higher Education Funding Council for England (1996–98).

Civil engineers and their work

Karl Kordina

I was born on 7 August 1919 in Vienna to the public notary Dr Karl Kordina and his wife Ludowika. At that time my father had, however, already been dead for three months; he died from injuries received in World War I. As the family had been deprived of their property through expropriation and inflation, my mother had to accept a teacher's job to provide for our livelihood.

Following the Abitur examination, which I took in 1937 at the traditional grammar school Zu den Schotten in Vienna, which had a humanist bent, I had three months training as a 'one-year volunteer' with the Austrian Army. In January 1938 I began studying civil engineering at the Technical University of Vienna, from which I graduated in 1941 with the main diploma examination.

During my time at university I was called up for front-line duty but resumed my studies after being wounded. Following a serious injury and conflicts with a political National Socialist officer, I passed the final years of the war as a lance-corporal until taken prisoner in North Germany in 1945.

In the autumn of that year I returned to civilian life and gained my first professional experience in the engineering departments of two medium-sized construction firms and in an engineering consultants' office.

In 1953, I started work as a member of the scientific staff at the Institut für Massivbau (Institute for Concrete Structures), of the Technical University of Munich under the direction of Prof. Dr.-Ing. H. Rüsch, where I also took my doctor's degree with a thesis on the load-bearing behaviour of slender reinforced concrete columns subjected to short and long-term loads. While working under Professor Rüsch, I had the opportunity to deal with complex practical problems and I was also concerned with highly interesting research projects. This allowed me to lay the foundations and expand my basic scientific knowledge. At the same time, I became increasingly involved in teaching practice and was encouraged to do my own scientific work and publish it. I was above all very lucky to have been introduced by Professor Rüsch to topical problems in research and to have been shown how to manage my own research projects.

In 1959 I was offered the chair of materials science at the Technical University of Braunschweig; in 1978, I was transferred to the chair of design of reinforced and prestressed structures. The federal state of Lower Saxony provided me with the opportunity to set up a comparatively large institute including a materials testing laboratory. The buildings, for which I provided most of the proposals myself, were to accommodate a highly efficient department for mechanical technology and building material and component testing, and also a separate department for testing fire-exposed building materials and structures. At the time, preventative fire engineering was only just starting to attract scientific attention. I, for my part, had first taken interest in these questions during my earlier work in industry, and my predecessor at the institute, Professor Dr.-Ing. habil. Dr.-Ing. E.h. Theodor Kristen had already prepared the ground in the field of fire protection (though with extremely inadequate testing installations). This enabled me to continue his efforts with much improved facilities.

The first scientific investigations into the fire performance of building materials and members took place around 1960, at a time when only a limited number of simple tests on reinforced concrete beams and slabs subjected to bending, or on columns and masonry walls, were available. High-temperature testing on building materials had up to that time only been carried out under the most common load conditions for concrete, steel and timber. Installations that would have permitted more sophisticated testing were not available at the time.

I made it my task to extend the test programme and perform – in collaboration with members of our staff – fire tests on reinforced concrete slabs and beams, both simply and continuously supported, and also on columns, frames and plane load-bearing structures, and to design and provide the installations required for such testing. We similarly conceived and performed tests on the construction materials involved, in particular concrete, so as to be able to define the resistance and deformation behaviour at high temperatures. The results thus obtained provided the basis necessary to quantify and draw conclusions by theory and callculation on the fire performance of structural members. We were greatly aided by the establishment by the Deutsche Forschungsgemeinschaft (German research association) of a special research department for the behaviour of structural elements exposed to fire, of which I was spokesman for over 15 years.

Our work met with considerable response, both nationally and internationally. Issues raised were taken up, thus launching intensive research activities on the international scene as well, which in turn led to comprehensive standardization of regulations concerning preventative structural fire design.

In the past few years much work has gone into the development of analytical models to describe fire histories calibrated on the basis of large-scale fire tests, e.g. those made in connection with the EUREKA tunnel fires in the north of Norway.

The work performed at our institute on the behaviour of fire-exposed building materials and components has greatly influenced, and contributed to, German standardization regulations and also efforts directed at harmonizing fire regulations internationally. The Science Council appreciated the work done by assigning it a key function; acknowledgement can also be derived from the fact that a special research sector was later set up for the fire response of structural elements.

The official civil engineering materials testing institute (MPA) concentrates on the testing of building materials and components for their compliance with the properties set out in the relevant standards, and thus serves a public interest. The close collaboration between the materials testing institute and the university institute has led to a fruitful relationship, and indeed the two bodies share facilities in common, and staff with special experience are available to both institutions.

In 1959, at the beginning of my Braunschweig activities, the institute staff comprised a total of 35 persons. Fifteen years later, their number had grown to about 200. Since 1972, the institute has been organized as a joint institute, first with two full professors and later with four professors. About 100 members of staff are employed in teaching and research, while administrative work and the materials testing institute account for a further 100. The wide range of issues in connection with materials and component testing, and also the extensive work flowing from fire engineering, were stimulus enough for us to embark on a correspondingly large number of research activities and projects in collaboration with the scientific staff. The result was a large number of dissertations, a number of habilitation theses and countless reports and publications. The list of my own publications alone covers more than 300 titles, which, it should be added, have for the most part been written together with staff members. Many of my earlier staff members have in the meantime assumed professorships or senior positions in administration and the building industry.

In my own research activities I have concentrated on the behaviour of structural members under fire, but also on the structural response of reinforced-concrete columns and frames, the stability of shell structures, problems and their solutions in the field of segmental structure prestressing, external prestressing and safety theory. As regards building materials, I have been concerned with such aspects as corrosion and surface damage of concrete, shrinkage and creep of concrete, and investigations into the problem of alkali reactivity and the consequences for the load-bearing capacity of components.

After 1960 I was appointed to a number of bodies of the German Committee for Reinforced Concrete, and I became involved in the development of reinforced concrete standards and the German DIN 4102 Standard. I am also involved in trying to safeguard German interests in leading bodies in the international arena.

Since my retirement, much of my time has been devoted to organizing and improving the Materialforschungs- und Prüfungsanstalt für Bauwesen in Leipzig, in my capacity as scientific director. The aim was to bring into line with current requirements an institution which has an important role to play for the region of Leipzig and also to create a link between the local universities and the building industry.

When asked today which of my achievements during my professional life I regard as worth mentioning, I would refer to the stimulus I have given for a large number of new research activities and developments, and also my endeavours to offer many young scientists an opportunity for qualification.

Karl Kordina

Univ. Prof. em. Dr. Ing. Dr. Ing. E. h.

Born in 1919 in Vienna, Austria.

Graduated from the Technical University of Vienna in 1941 and after working in two construction firms became, in 1953, a staff member at the Institut für Massivbau at the Technical University of Munich. In 1956 he gained his doctor's degree from the Technical University of Munich and in 1959 became Professor of Material Science and Reinforced Concrete Structuress at the Technical University of Braunschweig and head of the homonymous institute.

Between 1972 and 1986 he was head of the special research department concerned with the behaviour of structural elements exposed to fire sponsored by Deutsche Forschungsgemeinschaft.

In 1979 he recived a doctorate from the Technical University Bochum. He has written more than 350 publications concerned mainly with design of structures and the behaviour of materials and structures under fire.

The influence of history and opportunity on a professional career

Sam Thorburn

I was born in the City of Glasgow, Scotland, on 15 October 1930, within a district occupied by manufacturing industries. An impression of the life and style of the district at that period is given by the Figure overleaf. Four-storey masonry tenement buildings, shops at street level, electric tramcars and cobble stone streets dominated the scene. The juxtaposition of heavy industry and the tenemental dwellings was a common feature of the city environment. The only physical relief from this somewhat harsh industrial environment was the existence of extensive public parks which the city had prudently provided for the health and welfare of the inhabitants. Perhaps early evidence of the preservation of green belt within industrial cities.

The City of Glasgow has, however, an ancient history which shaped both the character of the city and its residents, and it would seem remiss of me to consider only the influence of recent history on my ancestors. The simple account which follows, however, is only intended to convey a glimpse of a rich tapestry of the lives of the early citizens of Glasgow.

Civilization first embraced the relatively few dwellers within the Glasgow district about 80 AD. Agricola, the Roman Governor of Britain, 78–86 AD, advanced into Scotland about 80 AD and penetrated as far north as Perth. Urbicus, the Roman Governor of Britain after Agricola, built a wall across Scotland about 140 AD. It was 37 miles long, with small forts at two-mile intervals, and was positioned a few miles north of Glasgow. The wall was built to confine the barbarous Picts to the lands north of the wall. This military defence strategy also protected the citizens of Glasgow and their development as a people was heavily influenced by Roman culture. About 200 AD the Romans withdrew south to the protection of Hadrian's Wall and left the people of Glasgow to continue a simple rural existence.

Three hundred and fifty years passed before the next historically important event took place with the advent of St Kentigern about 550 AD. We know that he was born in Culross, Fife, about 530 AD, the Son of Eugenius, third King of Scotland and Thametis, daughter of Lothus, King of Picts and was educated and trained as a priest of the Celtic Church by St Serf at the Culross Monastery.

GLASGOW CITY IN THE 1930S.

St Kentigern travelled to Glasgow and established a church within an area of green hills overlooking a crystal clear river, now known as the River Clyde. Here that he was visited by St Columba and he founded his church within the 'beloved green place' (Glasgu), hallowed by this visit. The church building may have been similar to the simple structures of wood combining church and cell favoured by St Columba and built in Ireland. Kentigern lived an ascetic and holy life until his death. He was canonized and became, on 13 January 603 AD, the patron saint of Glasgow.

The next historically-important event took place 500 years later, when Bishop Jocelyn was placed in charge of the church in Glasgow in 1175 AD. He was granted the right to form a Burgh of Regality by William the Lion, sometime between 1175 and 1178 AD and to hold a fair in 1189 AD for eight days from 6 July. Glasgow was unlike other burghs in that the common lands were held, not from the Crown or from a baron, but from the Bishop, and as their feudal over-lord the Bishop exercised domination over the people, particularly in the matter of local government affairs.

Glasgow was made in a Royal Burgh in 1611 by James VI, and this improve-ment in rank and privileges was confirmed in 1636 under a charter of King Charles I. The lands and privileges were thereafter held directly from the sover-eign and the taxes were paid to the treasury instead of, as formerly, to the Bishop. Despite this, the inhabitants, or rather the burgesses, had very little civic liberty. Not until 1690 can the city be said to have become absolutely free. Under a char-ter of William and Mary in that year, complete control in choosing their own magistrates and other officers for the management of the city was granted unre-servedly (as in all other Royal Burghs) to the free inhabitants of Glasgow.

The industrial revolution altered radically the nature of commerce in the City of Glasgow. By the beginning of the nineteenth century there were established numerous manufacturing industries, including iron and engineering works, pot-tery and glass works, chemical works, tar works, dye works and paper works. The St Rollox chemical works at Sighthill in Glasgow was formerly the largest of its kind in Europe.

The community within which I was born was dependent largely on heavy industry and, in particular, on a works known as Parkhead Forge, owned by the Beardmore family. The works covered an area of over 40 acres and employed about 4000 workers in the manufacture of armour plates, bridge beams, steam engine rods and pistons, propeller shafts for ships and plates for boilers. The com-pany was a major employer, and the district developed essentially in response to the demand for skilled workers.

Our family at this time possessed considerable steelmaking and forging skills, and some were employed on a specialist contract basis, employing their own work-men. They operated a 500–ton steam hammer to forge the red-hot metal and some worked with the furnaces which smelted Swedish pig iron.

The lives of people living and working within this district of Glasgow were bound by this industrial history, and their characters were shaped for at least two

57

generations. After closure of the works and subsequent dereliction of the site, the local community sought employment elsewhere and today a major shopping centre exists, known as The Forge.

About 1938, I developed a desire to be a civil and structural engineer when visiting relatives in Thames Ditton, Surrey, who owned a building firm. Although friends received secondary education at schools in Glasgow, I was educated at Hamilton Academy, a county school founded in 1588, which I left at 16 years of age, with the award of the Senior Leaving Certificate of the Scottish Education Department. The requirement to travel to and from the county town of Hamilton each weekday and spend most of the time away from home assisted the development of independence and the ability to cope with situations. Circumstances prevented full-time study and I worked under a training agreement with a civil engineering contractor based in Motherwell, an industrial steel-making town and studied during the evening at the Royal Technical College, Glasgow. The origin of the college, now the University of Strathclyde, was in January 1796, when John Anderson, Professor of Natural Philosophy at the University of Glasgow, left provisions in his will for the founding of a second university as a place of useful learning and liberality of sentiment.

Inter-disciplinary teaching of mechanical and electrical engineering subjects was a feature of the course of studies. Fitness of purpose for employment in industry in the west of Scotland was the clear objective. Although the physical and mental effort to sustain this form of learning was considerable, full-time employment made me aware of the relevance of the knowledge being taught at evening classes. This form of education and training extended my competence beyond knowledge gained at the Royal Technical College. I was seconded to Colvilles Ltd, steelmakers, in 1948, and spent ten years in the steel industry, continuing, therefore, quite by chance the family involvement in the iron and steel-making industry. Working within such a demanding environment in physical terms during those early formative years forged a resilience and determination which served me well in later years. This decade shaped my career, and the necessity to complement practice with full-scale research forged a pattern of philosophical thought which pervaded my 50 years in the construction industry.

During the period spent with Colvilles I carried out research into heat transfer in furnace foundations and high temperature effects on concrete, which helped me to satisfy an innate curiosity. Towards the end of the decade of involvement with the steel industry I became interested in soil mechanics and foundation engineering, but only in the context of structural engineering.

This interest was generated by involvement with the major civil and structural engineering works required for the redevelopment of the various steelworks owned by Colvilles, such as earthworks, roadworks, bridges, deep basements, rock and soil tunnelling, shallow and deep foundations, piling works, and river protection works. How fortunate I was to be presented with such opportunities for the early development of my professional career.

This wealth of experience was further extended by involvement with specialist construction companies between 1958 and 1966. These specialist firms were divisions of major construction companies and had been formed to carry out contract work in the specialist sector of soil mechanics and foundation engineering.

My wider interests in engineering prompted the formation in 1966 of a consultancy in civil and structural engineering, and it has developed over the past 32 years into an international firm employing about 450 persons.

The opportunities for research and monitoring the real behaviour of structures during the period 1958 to 1994 provided valuable information which I incorporated in over 60 technical papers and articles. An invitation by Dr Keith Whittles resulted in a book on underpinning and retention which was derived from my desire to transfer specialist knowledge for the benefit of young engineers.

I was commissioned in 1976 to commence the conceptual design of the new Glasgow Rangers Football Stadium at Ibrox, and in 1985 published an article in the *Journal of the Institution of Structural Engineers* on the design and construction of the stadium. This led to an invitation by the Sports Council in 1986 to present a paper at a conference at Harrogate, Yorkshire, on structures for new stadia developments. This invitation was repeated in 1990 when I was requested to present my thoughts on sports stadia after Hillsborough. The tragedy which occurred at Hillsborough Football Stadium during a Cup semi-final match on 15 April 1989, and which led to the deaths of 96 spectators, began an involvement with safety at sports grounds which still continues. I was appointed to the Board of the Football Licensing Authority in 1992 and was chairman of the task group responsible for the 1997 revision of the *Guide to Safety at Sports Grounds*.

My wide experience of the design and construction of heavy structures and their foundations led to chairmanships of technical committees of the Institutions of Civil and Structural Engineers. Involvement with Institution affairs, although time-consuming to persons following a busy professional career, is essential for the continuance of a profession and should be regarded as an honour to serve and not a burden to carry.

In order to progress knowledge regarding the real behaviour of structures in contact with ground, I chaired a national committee of the Institution of Structural Engineers, and a state-of-the-art report on soil-structure interaction was published in 1978. This report stimulated fresh interest in the subject and encouraged research into the real behaviour of structures. Subsequent research directed by me in conjunction with the Building Research Establishment is recorded in several technical publications. In 1985 it was deemed prudent to review the relevance of the 1978 document to current practice and the need for revision and extension was identified.

The Institution of Structural Engineers, with the cooperation of the Institution of Civil Engineers, and the International Association of Bridges & Structural Engineering, initiated the preparation of a comprehensive guidance document covering most types of structures, and I served as chairman of the review panel, and edited and contributed to the 1988 document on soil-structure interaction.

If I stand accused of over-zealousness in promoting an interest in soil-structure interaction, I accept that criticism. However, I consider it essential to consider the global nature of structural engineering problems with the necessity to consider the interaction between ground and structure.

In March 1981 I was appointed by the chief civil engineer of British Rail (Eastern Region) to investigate major problems being experienced with the construction of foundations for bridges associated with the East Coal Main Line, including the Ryther Viaduct. A complete account of these problems and their solutions has been presented by British Rail in the *Proceedings of the Institution of Civil Engineers*, Part 1, 1985 (Paper No 8821). The resolution of these difficulties led to my appointment by British Rail as an advisor on the design and construction of bridge foundations. I was appointed also as an assessor for the Dartford River Crossing Project in January 1989 because of long experience of bridges and their foundations. These two appointments arose mainly from my knowledge of practical methods and techniques of construction gained from first-hand site experience.

Earlier involvement with the steel industry also necessitated the development of solutions to the problems of industrial wastes disposal, and subsequent experience of treatment of ground underlain by household and other wastes resulted in an invitation to be chairman of a conference on building on marginal and derelict land, held in Glasgow in 1986. The conference dealt with landfill sites and received the support of the Scottish Office. Since 1972 I was involved with the design, construction and supervision of landfill sites regulated by the Mines & Quarries (Tips) Act 1969. Once again, my career had changed direction and opportunity was given to become involved with engineering problems of a different nature.

Experience of old mineral workings within the central belt of Scotland resulted in commissions to design special foundations for medium and high-rise buildings to accommodate the ground strains associated with mining subsidence. This experience included consolidation of old mineral workings by grouting techniques. This involvement resulted from friendships and discussions with geologists and mining engineers during earlier years, which created both interest in the subject and the acquisition of knowledge.

Practical research has occupied much of my career and I was given the opportunity to investigate the fatigue of concrete and steel on behalf of the Departments of Transport and of Energy. In 1988 my contributions to technical committees of these Departments resulted in a commission from British Petroleum (BP) to prepare a report on the means of securing tethered buoyant platforms in the harsh regime of the North Sea. I innovated and developed a seabed anchorage system for tethered buoyant platforms, which was patented by BP International Ltd in 1980. Subsequently, I was commissioned to prepare a reference document on the design of foundations for offshore structures, financed by the Department of Energy.

Once again my career changed direction and I received a commission in 1987 to survey the blast wave damage caused by severe explosions at two major chemical plants at Grangemouth, Scotland and Antwerp, Belgium, which involved analysis of the structural responses of various types of buildings to the peak overpressures. This opportunity to become involved with severe explosions and their effects was unexpected and is indicative of the inability to predict one's future. The twists and turns of fate have certainly created the fabric which has formed my life.

In 1988, advice was sought by the Home Office in relation to the protection of traditional buildings against explosions using innovatory concepts and in the development and testing of blast-resistant structures. In June 1989, the opportunity was given to participate in the tests, known as Misers Gold, at White Sands Missile Range, New Mexico. The photograph shows the cloud of desert dust created by the simulation of a blast equivalent to a 4 kiloton nuclear device. The explosive mixture was contained within a fibreglass dome measuring 13 metres in diameter at the base and 11 metres in height.

The experience of working in a high-risk environment and close to a lethal explosive mixture was unique. The final experience of witnessing the explosion while standing in the open at a safe distance, together with the physical feeling as the blast wave passed through the spectators, is difficult to describe. Quite by chance, I subsequently acted as specialist advisor to the Defence Committee of the House of Commons.

In 1982, I was approached by the then president of the Institution of Civil Engineers to monitor events in Europe and, in particular, decisions taken by the Commission and the European Parliament which would affect members of the civil engineering profession in the UK. This involvement with European affairs extended over a period of ten years and resulted in a series of papers and articles. The Institution of Civil Engineers awarded the Rear Admiral John Garth Watson medal to me in recognition of services in respect of European affairs.

A tragic accident occurred in 1992 in Bastia, Corsica, which caused the deaths of 17 spectators and injuries to over 2000 persons. The collapse of the temporary demountable structure which had been erected for a football match in Corsica initiated questions in the House of Commons and actions by the then Minister for the Environment. I was appointed chairman of the technical committee responsible for the preparation of guidance on the procurement, design and use of temporary demountable structures. Subsequent to publication of guidance by the Institution of Structural Engineers in 1995, a monitoring group was formed incorporating industry and local authority representatives and I was appointed chairman.

Numerous appointments have resulted from a long involvement with professional affairs and the work of important technical committees. The need to modernize the British Standards Institution resulted in the author being appointed a member of the Technical Sector Board for Building and Civil Engineering and Chairman of the BSI QA Certification Advisory Council for the service sector.

I have attempted to highlight events which shaped my career, although much of the detail has been omitted for the sake of brevity. I am indebted to many mentors and colleagues for my progress in life. In retrospect, a career in engineering comprises opportunities, disappointments, successes, periods of quiet development and times of great pressure to complete tasks in a relatively short period of time. The avoidance of errors requires more than competence and concentration when insufficient time is available for calm and measured decisions.

Finally, I was elected president of the Institution of Structural Engineers for the session 1997–98. This great honour was most unexpected, as was the honour of being made an Officer of the Most Excellent Order of the British Empire (OBE) in June 1987. The earlier election in March 1984 to the Royal Academy of Engineering completes this commentary of a life beginning in the streets of Glasgow. It has been a rich and rewarding experience, and my final comment is that one does not choose one's time and place of birth. Fate has the final say in the path of destiny. The greatest pleasure I have experienced was my appointment as a visiting professor in structural engineering at the University of Strathclyde in 1985. That my Alma Mater would give me the opportunity to assist in the education and training of young persons completed my aspirations.

Sam Thorburn, OBE

OBE, DSc, FREng, FIStrucE, FICE, FASCE

Sam Thorburn has been a Visiting Professor to the Structural Engineering Department of the University of Strathclyde since 1985. He is also a Trustee of the University. A Fellow of the Royal Academy of Engineering since 1984, he has been appointed as Chairman of many national and international committees and was Chairman of the committee responsible for the new *Guide to Safety at Sports Grounds*. Author and co-author of over 60 professional papers and articles, the latter mainly on the European Community. He is a member of the Technical Sector Board for Building & Civil Engineering and of the Football Licensing Authority, created as a result of the Hillsborough tragedy in 1989.

In 1966, he founded the consultancy now known as Thorburn Colquhoun, an international firm, and now part of the URS Corporation, USA. He was elected President of the Institution of Structural Engineers for the session 1997–1998.

Concrete: the indispensable material

Gunnar Idorn

Origins

At the age of 15 in 1935 I was picked out by my father and the schoolmaster in the fishing town of Gilleleje, North Zealand, Denmark, as fit for higher education, with a mathematical and scientific bent. I was admitted to the Technical University of Denmark in 1938 and graduated as a civil engineer (M.Sc.) in 1943. For my graduation project in coastal and port engineering I designed an extension of the Gilleleje fishing and packet-trade harbour. This choice of career evolved from my boyhood's work and free time at the harbour. In addition, I was attracted to mathematics and science, and interested in reading about technology and civil engineering which were held in high public regard as being indispensable for social progress . World War II in occupied Denmark 1940–45 passed with a preliminary job from November 1943 with a land surveyor's firm. This alternated with, and from late 1944 was replaced by underground resistance until liberation on May 5, 1945. November of the same year offered an opening as coastal and port engineer in West Jutland for the Board of Maritime Works, and this set the theme for my subsequent professional life.

The west coast of Jutland, 1945

Denmark was spared major devastation during the war because it was not made a battlefield. Through daily intelligence contact with the allied forces' headquarters in London, air-raids and underground sabotage were planned and carried out on selected targets. Socially, however, the country was worn down, and industry and power production, fuel and commodity supply were at the lowest ebb. The remote coastal regions of West Jutland had been made part of the German 'Atlantic wall'. Hence, maintenance had been reduced for the few small fishing ports and for sluices, many coast-protection groynes and single breakwater jetties, and dykes of 10–50 km length which protected the lowlands behind against flooding. With liberation came a tremendous need for reconstruction. At the same time the 20–

30% unemployment rate made labour-intensive construction programmes politically desirable. Although government allocations for such programmes were limited by the financial situation, the primary barrier against renovation was the dilapidated condition of the pre-war machinery and heavy vehicle depots, the lack of spare parts and the strict rationing of fuel. Our frustrations over these circumstances were countered by the underlying strong motivation generated by having won our liberty again. We could see tangible progress almost from month to month in daily life for the local people. Also, the will to succeed on the part of crews and foremen helped, despite the conditions, in the much needed production of high quality concrete.

Concrete technology innovation

In 1948 the American Marshall Aid Programme initiated a remarkable increase in general industrial capacity and building and construction capacity in Denmark, as in all of western Europe. I was appointed resident engineer for reconstruction of the Hanstholm Western Jetty, for which the aid programme supplied new heavy vehicles, a screening plant for aggregate production with beach sand and gravel as the resource, a concrete batch and mixer plant, and a concrete pump. The site was at the northwestern point of the west coast. The west jetty of a planned harbour had been near completion in 1940 when the German Army occupied the site to construct a long-range cannon fortification with a 12,000-man occupation force. The 200 local people of the harbour workforce and local fishermen families were forcibly evacuated, and the unfinished jetty left to deteriorate. Figure 1 illustrates the restoration work, where steel bungs were used to encapsulate the broken-down pre-war caissons in new mass concrete.

FIGURE 1 *Renovation of pre-war western jetty, Hanstholm, by placing mass concrete between new bungs and old semi-collapsed concrete caissons.*

FIGURE 2 *A gale over the western jetty, Hanstholm. Reconstruction work on the jetty was possible on only about 70 days per year.*

The following four years saw intense efforts to learn by trial and error how to produce concrete with the new, mechanized technology at a production capacity never before available in Denmark, and to comply with demands for high-quality concrete. Figure 2 illustrates why the exposure to wind and waves restricted castings on the jetty to 70 days a year, which meant that 2–3 day and night continuous concrete making and placement often took place. We were forced to carry out practical research into how to maintain the pumpability of fresh concrete, and the possible cracking effects of heat development during early curing of concrete, in order to meet this new requirement for concrete engineering.[1]

Frequent and often prolonged periods of bad weather allowed time to study more basic concrete technology and the excellent past records of pre-war construction works by preceding coastal engineers.[2] I re-established their inspections of the service condition of older concrete blocks in groynes and jetties along about 200 km of coastline. Figure 3 shows plain concrete blocks in the south jetty, Hvide Sande, unscathed after about 80 years of severe exposure, illustrating the excellent craftsmanship possessed by the operating crews and engineers of bygone days. There were other older blocks with patterns of surface map-cracking, which I also found in many other coastal structures.

1. Idorn, G.M. *Concrete on the West Coast of Jutland – Part 1*, Danish National Institute of Building Research and the Academy of Technical Sciences, Committee on Alkali Reactions in Concrete, Copenhagen, 57 pp. Progress Report B1. 1958.
2. Idorn, G.M. *Concrete on the West Coast of Jutland – Part 2*, Danish National Institute of Building Research and the Academy of Technical Sciences, Committee on Alkali Reactions in Concrete, Copenhagen, 54 pp. Progress Report B2. 1958.

FIGURE 3 *Mass concrete blocks at Hvide Sande cast in 1915, photographed in 1993.*

At the same time similar, more severe, surface map-cracking in the piers of highway bridges in north-west Jutland was found to have been caused by alkali-silica reaction. This exciting dual experience of new concrete technology and the durability and deterioration of older concrete stimulated my personal interest in learning more through research, and led me in 1952 to accept an invitation to join the new Danish National Building Research Institute (SBI).

A new era of concrete research in Denmark

The transfer was tempting for several reasons. Moving back from the remoteness of the west coast to the metropolitan area offered access to the new influx of applied science in cement and concrete research from the USA and UK, while the political demands for new house building and infrastructure construction brought concrete high on the priority list for public services, and therewith abundant financing for research. Moreover, the older generation in the higher management networks in the building and civil engineering sectors, who mostly sought a 'back to pre-war normalcy', could not match the younger engineers' determination and energy in the quest for new knowledge and technology.

Winter concreting

The shortage of housing and the contemporary baby boom made 'build the whole year' the slogan for a long-term building research programme. The shortage of

67

brick production capacity and of trained masons was an incentive to make concrete production possible during the traditional November to April cessation of building and construction. The SBI team combined new basic knowledge of the physical properties of hardened cement paste from research in the USA[3] with British research[4] on the impact of the kinetics of cement hydration on the temperature rise during concrete curing. In less than a year this new knowledge was adapted to a practical method for monitoring the rate of concrete strength development at low temperatures.[5, 6] Readings of the temperature change in the actual concrete enabled calculation of the rate of maturity development and hence of the necessary time before removal of formwork, the need to supply external heating on site, etc. This was a more innovative approach than anywhere else at that time. Implementation of the new technology by government regulations and comprehensive training courses for engineers and concrete work crews made the modest research costs pay off very well, and stimulated SBI's further work in new concrete technology such as general quality control, statistic evaluation of control test data, etc.

Alkali-silica reaction (ASR)

In 1951, visitors from SBI to the USA brought back news of the discovery that siliceous aggregates could react deleteriously with alkalis from the cement in concrete, and shortly afterwards the severe cracking in the highway bridges in northwest Jutland was found to be caused by ASR.[7] When I joined SBI early in 1953, public authorities and the cement industry had decided to create a joint investigation programme because ordinary gravel and sand in Denmark were undoubtedly reactive, and Danish Portland cement was of sufficient alkali content to make the reaction a general threat to the necessary reliance on concrete as the primary reconstruction material. The government, the cement and aggregate industries, contractors and consultants financed the project with the Academy of Technical Sciences and SBI as management partners. I became one of a small team of civil engineers who then needed to educate themselves in basic cement and concrete chemistry, mineralogy and geology, and to adopt and invent new investigation methods for assessing the nature and seriousness of the ASR problem, and also

3. Powers, T.C. and Brownyard, T.L. *Studies of the physical properties of hardened Portland cement paste.* Research Laboratories of the Portland Cement Association, Bulletin 22 (reprint compilation). Chicago, March 1948. 1
4. Saul, A.G.A. Principles underlying the steam curing of concrete at atmospheric pressure. *Mag. Concr. Res.*, **6**, 127–35. London, March 1951.
5. Nerenst, P., Rastrup, E. and Idorn, G.M. *Betonstøbning om vinteren*, Teknisk Forlag, Copenhagen, 2nd rev. ed., 91 pp, in Danish. Series: SBI Anvisning nr. 17, 1958.
6. Rastrup, E. Discussion to reference 63. Session BII. *Proceedings of the RILEM Symposium, Winter Concreting.* 35–50. Danish National Institute of Building Research, Special Report. Copenhagen, 1956.
7. Nerenst, P. *Betonteknologiske studier i USA – Rapport over ECA studierejse 14. November 1950-15. februar 1951*, Danish National Institute of Building Research, Copenhagen, 88 pp, in Danish. Series: SBI-studie nr. 7, 1952.

eventually to find the means to overcome it. In two summer seasons we then thoroughly 'mapped' the occurrence of ASR-affected structures and collected samples of aggregates and cement. We found that although there were enough concrete structures with severe deterioration to justify the long-term research, there were many more with no evidence of deleterious effects; in other words, the reaction could also assume a harmless course in field concrete. That necessitated thorough, explanatory research and led to a fruitful exchange with ASR research groups in the USA, Australia and England where studies of ASR were underway.

In our programme we emphasized the development of a diagnostic methodology for assessing the causes of deterioration, because deleterious reactions other than ASR also occurred, notably chemical reaction caused by the sulphate in sea water, and damage due to freezing and thawing. Each of these kinds of deterioration were found to leave their own fingerprints in the concrete. We undertook systematic analysis of cores drilled out of field concrete, and microscopic examination of thin sections prepared from the cores. Concurrently, geologists examined and recorded the origins and compositions of aggregates used in Denmark, and about 4000 mortar bars were cast modelling the influence of the various parameters which were believed to influence the course of the reactions. The studies were described in a series of 22 progress reports issued from 1956 to 1967. I was the author of eight of these, and the principal author of a general report on chemical reactions involving aggregates.[8] I further compiled the field core and thin section examinations in a detailed diagnostic methodology for field concrete deterioration.[9] For this achievement I was awarded the D.Sc. at the Technical University of Denmark.

While retaining my full-time job as senior research officer at SBI, I agreed in 1957 to establish a materials technology department for a new Bachelor of Engineering school, the Danish Engineering Academy. In the summer holidays of 1958 I wrote a new 164-page textbook on cement, the first ever issued in Denmark, and several shorter books on concrete technology. This transfer of science and technology to education was a rewarding experience.[10]

The Concrete Research Laboratory, Karlstrup (BFL)

In 1960, Aalborg Portland-Cement-Fabrikker A/S decided to build and operate a concrete research laboratory as a service for its customers, especially consulting engineers, contractors and the new precast concrete industry. The company produced about 90% of the cement used in Denmark, was a world leading exporter

8. Bredsdorff, Per, Idorn, G.M., Kjxr, Alice, Plum, Niels Munk and Poulsen, Ervin: Chemical Reactions Involving Aggregate. *Proceedings of the 4th International Symposium on the Chemistry of Cement, Washington 1960, Vol. 2*, National Bureau of Standards, Washington, 749–806. (with bibl.) National Bureau of Standards Monograph 43, **2**,1962.
9. Idorn, G.M. *Durability of Concrete Structures in Denmark*. 208 pp. Teknisk Forlag, Copenhagen, 1967.
10. Idorn, G.M. *Grundtraek af Betonteknologien 1. Cement*. 164 pp. Polyteknisk Forlag, Copenhagen, 1958.

of white cement and, was by ownership associated with F.L. Smidth & Co, who produced about 30% of the cement plants in the world. I was engaged to set up and lead the new laboratory, known in Denmark as BFL-Karlstrup.

To me the new job was a godsend. SBI's emphasis on concrete research over more than a decade could not continue to match other priority building research issues, especially when the cement industry announced its willingness to take over the financial and management obligations. My qualifications satisfied the need to introduce more basic science in concrete research, which I had already worked towards through articles, committee work, etc. The prospect of being influential on a broader research platform with secure long-term financial support was an offer not to be refused. True, I had no university research training or education in commercial management, but there were no competitors for the position with such qualifications.

From commencement until closure in 1976, BFL was synonymous with an entirely new cement and concrete research epoch in Denmark with a staff of researchers, technicians and service personnel rising to 45 people. A detailed survey of its development, strategy and accomplishments is given in *Concrete Progress*[11]. BFL won recognition in Denmark and also in the international cement industry and academic research communities; we appeared frequently as invited speakers at conferences and authors in periodicals.

A pioneering technology transfer of our long-term research on the effects of cement hydration during site and pre-cast concrete production resulted in the maturity computer, an instrument that monitored the curing of concrete at elevated or low temperatures. Figure 4 shows high-strength concrete achieved through the use of a special low-porosity cement, alternatively attained by polymer or sulphur impregnation. Concurrently, we adopted new instrumentation such as X-ray diffractometry, isometric calorimetry, SEMEX, electrodynamic vibration, EDP, etc. From about 1972, when we were promoted to be the R&D department of the company, we adopted the industrial research management system which we acquired through membership of EIRMA (European Industrial Research Management Association). Figure 5 depicts the principle of the strategy and management system which directed our various operations.

In the course of 1974, I observed that we could not rely on the continued financial grant from our single source, now renamed Aalborg Portland A/S. In the wake of the oil embargo in 1973 and the approaching saturation of the housing and infrastructure markets, cement consumption plunged almost 50%. This coincided with a trend to identify new targets for capital investors, away from smoke-stack industries, to electronic, high-technology pioneering enterprises. Hence, on a sad day in April 1976, the decision was announced to close BFL-Karlstrup and the five smallest of the six Danish cement plants. The fraction of the annual sales profit which in the past 16 years had been allotted to research was

11. Idorn, G.M. *Concrete Progress, from antiquity to the third millennium.* 359 pp. Thomas Telford Publishers, London, 1997.

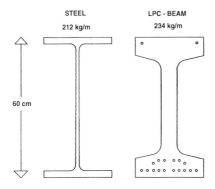

FIGURE 4 *High-strength, prestressed concrete beam: height and load-bearing capacity equal to steel-profile beam of approximately same weight. Special development with low-porosity cement (LPC) at BFL, 1975.*

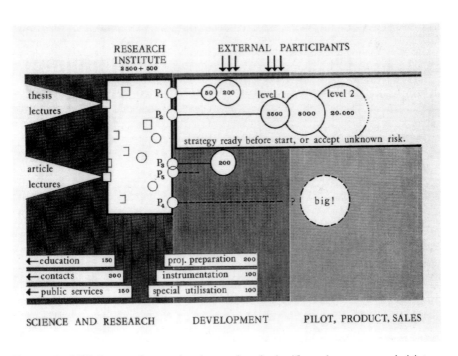

FIGURE 5 *BFL's integrated approach to its complex of scientific, explanatory research, joint venture implementation projects, outreach communications, and management of R&D operations (Figures are in thousand Danish kroner).*

71

said to have vanished. In other words, in the drastic company down-sizing, research was considered expendable. I had fought vigorously and proved that we were a cost-effective asset to the company and a superb agent for promotion, but to no avail. The long-term loss of a strong, national centre for advanced research and technology development was not considered in the short-term cost-cutting philosophy.

A fourth career avenue

Two subsequent years in the company as manager of overseas joint venture projects convinced me that a new career as an independent consultant would be an attractive challenge. It was not a step to take lightly, in view of the world-wide downsizing in the building materials and construction sector. However, I was well-known, and believed my recognition acquired at Aalborg Portland could be turned into a personal marketing asset.

Initially I promoted my skills in terms of the R&D management system, which had been a personal development at BFL, but no clients wanted to buy that. On the other hand, my first-hand understanding of the basic concept of conversion of chemical energy to mechanical work, first during the processing and next during the performance of concrete under different environmental circumstances, proved to be an important asset for advisory services in search of solutions to clients' specific problems. For the next eight years I became involved in the serious consequences of using obsolete 'cold regions' specifications for concrete making in hot Middle East countries, and in legal complications with standard specifications in countries where the occurrence of ASR had been unduly ignored. Progressive clients in North America and Europe sought my advice in promoting the use of cement in blends with fly ash or GGBS (ground, granulated blast-furnace slag). It was also a special pleasure as consultant for the construction of the Danish Faroe Bridges to contribute to the implementation of much of the previous research from BFL. I received calls for assistance in different parts of the world including North America, Africa, China, the Middle East and Europe, which enabled me to participate in international research conferences, maintain contacts with colleagues and institutes, and market my services.

Towards privileged maturity

My life as a single, international consultant turned out to become so time-consuming with about half the year spent away from home that in 1986 I sold it as a 'going enterprise' to the major Danish consulting firm Rambøll & Hannemann A/S, and after three years as director of this new G.M. Idorn Consult A/S subsidiary I resigned, approaching the age of 70.

The growth of consulting services in my field was the result of the new trend, especially in the UK and the USA, towards litigation in cases of failures in concrete structural design and early deterioration of field concrete. This required lengthy investigations to assess the nature and cause of failures. The consultant's impartiality and ability to communicate were crucial, both in reporting and at hearings. Concurrently, the fall in the general, social and financial prestige of building and construction investments and therewith of funding for progressive research had the consequence that the legal profession inadvertently became a major sponsor of research. In recent years this has increasingly resulted in the financing of competitive investigations to serve claimants or defenders' interests, but not progressive development of science or technology. This made me decide on a final personal investment, to publish an extensive survey of about half a century's professional life with concrete.

Concrete progress

The book, with its subtitle *from antiquity to the third millennium*, deals with the technology of concrete, the indispensable material for building and construction, in the social and economic history of mankind. In this context I describe the 16 years rise and eventual fall of BFL in Denmark, review almost 60 years of international research on ASR, and draw attention to the now increasing separation of academic and institutional research from realism and cost-effectiveness, i.e. from virtual leadership. Finally, I explain why, for the new millennium, global demography and resource issues – materials, knowledge, investment capital – require a new paradigm for concrete research with top-level, industrial leadership involved, and recognition of the dependence of socio-economy of the developing countries on substantial, cost-effective concrete progress. The means to realize this are, for those involved, far more powerful than what was available when I commenced in 1945. However, the need to accomplish and the hindrances to overcome are no less challenging than in those bygone days. This seems a good message to hand over to the concrete technology engineers and researchers of the future.

Gunnar M.Idorn

Gunnar Idorn was born in 1920. He graduated as a civil engineer (MSc) from the Technical University of Denmark in 1943and after national service in 1943 and 1945 he served with the Board of Maritime Works on the west coast of Jutland until 1953. From 1953 to 1960 he served with the Danish Building Research Institute and managed the field and laboratory investigations of some 600 concrete structures in a ten-year programme with focus on alkalisilica reactions. He introduced thin-section petrography as a means of studying chemical and physical reactions in concrete.

The DSc was awarded in 1967 and as lecturer at the Academy of Engineers, Dr Idorn wrote the first Danish textbook on cement for civil engineers. He was head of the Concrete Research Laboratory, Karlstrup, 1960–75 and initiated a series of international conferences on alkali-aggregate reactions. From 1978, G.M. Idorn Consult has been active in the international field, after 1986 as affiliate of the Danish consulting company, Ramboll & Hannemann A/S, with Dr Idorn as management advisor since 1989.

He is a member of the Danish Academy of Technical Sciences and recipient of the Ostenfeld Gold Medal. He was elected honorary member of the ACI in 1981; presented the honorary Raymond E. Davis lecture in 1983 and has been active in several ACI committees. In 1990, the ACI arranged a three-day symposium on concrete durability in his honour held in Toronto. He is a member of the Honorary Editorial Board of the journal *Cement and Concrete Research*, member of RILEM and the Danish Section of the French Civil Engineering Association.

Gunnar Idorn is the author and co-author of more than 200 technical and scientific articles on cement and concrete research and on planning and management of research. His dissertation *Durability of Concrete Structures in Denmark* (Teknisk Forlag, Copenhagen, 1967) pioneered the method of diagnosing field concrete deterioration by integrated field and thin-section investigations. His latest publication is *Concrete Progress,* (Thomas Telford, London, November 1997). It emphasises the socio-economic importance of concrete in building and construction from antiquity into the third milennium.

The joys and tribulations of innovation in civil engineering

Peter Head

I wanted to create structures from an early age. I lived in a typical South London suburban street with pebble-dashed semi-detached houses hiding long thin gardens. Beyond that neat hedge just over halfway down the manicured garden was a largely unseen world where huge earthmoving operations, deep excavations and major structures silently came and went. Beach holidays in Cornwall provided opportunities to explore dam building on a monumental scale. A day would be spent constructing a sea wall that would protect our small piece of Polzeath beach from inundation for at least 15 minutes after our neighbours had scuttled off up the cliff for a cream tea.

Then there was the winter of 1962/63 when I was 15, with month after month of hopeless igloo building, marvelling at the beauty of gnarled icicles and wondering at the avalanche risk of the neighbours' roof snow crashing down into our side passage and whether my sister would be there at the time.

Behind those neat suburban net curtains was any amount of passionate curiosity about what made the world tick. I needed to explore it for myself and began to piece together a picture which gave a very shy person some confidence. Most of this was an exploration of engineering principles but I neither knew nor cared. For me I somehow sensed that if I observed and understood these things I would be able to cope with life better.

I was learning for myself that this 'understanding' was fundamental to civilized life and I sensed that to be in control would perhaps lead me forward. This feeling was quickly reinforced in the Scouts where camp challenges of crossing rivers on aerial ropeways, coping with torrential floods, cooking in the open and building shelters in wild mountain glens became a regular feature of my life. This was also a steep learning curve for management skills because survival in the human Scout jungle depended on it.

Another major passion in my life at the age of 15, apart from fantasies about the untouchable girl that would cycle downhill past the house, was music. I sang in my church choir at the top of the hill every week and scaled great heights of achievement as a solo chorister, my first real success in life done entirely on my own. This gave me confidence to perform in public, as long as I was well practised, which has been very valuable in later life. But even more important was the link established through my love of music with the spiritual dimension of design. Music was miraculously designed and, through their artistry, composers such as Sibelius were capable of transporting me from my little suburban life to ecstatic journeys through the snowfields and forests of Finland, whenever I wanted to go.

Somehow I was lucky enough to come to understand, in that unshakeable way that one believes the laws of physics, that design is an art that connects and engages the human spirit through the senses and that engineering is fundamental to every aspect of human life. The successful combination of the two in improving quality of life is true innovation, and this was something I would not realize for another 25 years. However the seeds were sown in that suburban life of the late 1950s and early 'swinging sixties'.

Not surprisingly, bearing in mind my interest in design, I was tempted with architecture but Tiffin School and its academic emphasis together with my grandfather's engineering genes drove me down the maths, physics and chemistry route to a place at Imperial College to study civil engineering. I had no living influences in my life to follow this route and the school careers advisor had not even heard of the subject. I talked to no-one about it but it just seemed the best thing for me. Perhaps my writing this will help you to see that we all suffered the same uncertainty and worrying lack of confidence that we were doing the right thing. Trust your senses and go for it!

College between 1966 to 1969 started with a year of having a good time in South Kensington and then I met my adorable wife-to-be who inspired me to work extremely hard for the next two years. I left Imperial College to join Freeman Fox & Partners with a few facts that I quickly forgot, but with a priceless gift which I called ' an approach to problem solving' at the time. Professor John Burland has recently called it 'a habit of mind'. It was exactly what I needed, the tools to continue my exploration of the understanding of engineering.

I did not have a plan then as to where my career would take me and I do not now. I have followed my nose and instincts and, looking back, I have been very fortunate, but have always made the most of opportunities that came along. Another key element of successful innovation is the determination to succeed and this requires surprisingly large amounts of courage in the face of adversity, as we will see later.

I worked for Freeman Fox from 1969 to 1980. I joined them because they offered me a job and I was drawn to their reputation for innovative bridge engineering. A few months after I joined, their steel box girder bridges at Yarra in Australia and Milford Haven failed with loss of life. These events changed the course of bridge design worldwide and fundamentally influenced my career in more ways than I could ever have imagined at the time.

A complete reorganization of staff and projects was undertaken in the London Offices. Oleg Kerensky and Bill Brown were preoccupied with the official enquiries and Bernard Wex was asked to take over the completion of Milford Haven Bridge and the design and construction of Avonmouth Bridge, a major steel box girder bridge which had just started construction. The Merrison Committee was set up to steer these projects to completion. I was appointed as team leader for the checking, redesign and construction of Avonmouth Bridge which was an awesome challenge that occupied me flat out for three years. We were developing the rules, applying and correcting them, strengthening steel already erected, designing later sections and developing advanced erection techniques to save time and money. It was engineering at the absolute sharp end with huge steel boxes cantilevered 65 metres out over busy main roads with the knowledge of recent collapses fresh in everyone's mind (see photograph below). We measured every deflection and found we could predict them to within 10%. Up to that time deflections on other bridges were often double those calculated by simple means and I was told that you could not predict steel bridge behaviour accurately. We gained in confidence very quickly and had a brilliant team. Anthony Freeman worked closely with me as my opposite number at Fairfield Mabey, the contractor. Alan Rees was the site agent. Suhken Chatterjee led the Department of Transport's Merrison team and Srinivasan led the independent check team at Ove Arup. However the most important technical mentor was Walter Brown at Freeman Fox who, above all, gave me the confidence, understanding and computing tools to completely understand the structure in all its complex three-dimensional behaviour. These finite element tools were only just being developed and we were using the most advanced versions to model complete box girders.

Another aspect of true innovation that was central to the success of this extraordinary period was the seamless involvement of many engineering disciplines

M5 AVONMOUTH BRIDGE.

such as mechanical, naval architecture and manufacturing. Nowhere was this better illustrated than at seminars held at Cambridge University, where Professor John Dwight brought together eminent researchers and key practitioners to discuss and debate the latest findings in thin plate steel behaviour, much as had been done at the Royal Society during the industrial revolution. It was here on a cold snowy winter's day that I first met Professor Douglas Faulkner, an expert in ship design, who was to play a key role in influencing my later career.

I was now 25 years old and I had the satisfaction of seeing Avonmouth Bridge completed using an integrated team approach and the development from scratch of innovative design and construction technology. This was accomplished in the teeth of a huge backlash against the technology because of earlier failures. I had seen the simultaneous research, analysis and construction of a major structure – and had become the father of two fine children.

M5 AVONMOUTH BRIDGE.

Constant soul-searching, checking and cross-checking were an everyday fact of life in this brave new world of innovation. There were only disastrous precedents to fall back on. However it was still a relatively narrow field of expertise and I felt that I needed more direct hands-on site experience if I was to really make a major contribution. It takes a long time to learn the complex interplay between technology, specification, contract and execution.

I then spent six years on major construction sites learning the trade, and this was only just enough (photograph opposite). I learned as much about inefficiency and waste as I did about the value of real skills and experience. I was depressed to see these skills were not being nurtured or valued very often. I learned that successful construction is as much about human interaction as it is about technology,

FRIARTON BRIDGE, PERTH.

because the way the contract is managed determines how the project is executed and work is still so labour intensive.

Freeman Fox were somewhat burnt out and not receptive to my desire to move forward. I joined Maunsell in 1980 and, with a young family, Susan and I moved back to London to settle in Beckenham close to Maunsell's head office. I had read the Yarra Royal Commission Report but had not made the connection that Maunsell in Australia, as designers of the substructure of the bridge, had staff on site at the time of the collapse. Maunsell were in no way criticized but the staff were deeply affected and the practice suffered as a result. Their momentum and position as world leader in concrete bridge construction were being threatened by the French. However, David Lee and the late Dr Brian Richmond encouraged me to push forward with new ideas, and the chance to really try something new came in 1981.

Maunsell had been commissioned to study the provision of maintenance access to the steel plate girders of the A19 Tees Viaduct. Coincidentally, the Government's Transport and Road Research Laboratory had tried out a concept of bridge enclosure and shown that wrapping steel girders in a bridge like this with a plastic membrane can reduce corrosion to a negligible level. The innovative leap was to discover that if this membrane was a structural floor made of durable reinforced polymer material, providing an access platform, the solution would provide a cost-effective answer to the problem as long as the floor was not too expensive.

I remember vividly the frisson associated with the key steps in this trail of discovery. The first was a steel conference held at Teesside Polytechnic in 1981. My old friend Professor Douglas Faulkner was there and we travelled back to

Darlington on the train together afterwards, passing under the very bridge I was studying. I mentioned my new idea to him as we went underneath and in the noisy DMU he turned to me with a broad smile and said 'That is one of the best uses of glass reinforced polymer materials I can think of. We have been researching them for structural uses in minehunters at Dunfermline. You should go and read (the late) Charles Smith's papers. You really should pursue it!'

I went to the Institution of Marine Engineers the very next day and can remember the great excitement as the papers were laid out and I marvelled at the work done. From that day to this I have been convinced of the value of these materials in civil engineering. The manufacturing methods were too crude for low-cost construction however and so I set out to find a new concept. Suddenly a whole world of innovation opened up and the hunt was on with no time to lose. I discovered a process called pultrusion within days and invited a manufacturing expert, Ron Brown, to our office. He described the ability to make complex cellular forms and I knew we had the answer. I remember going to talk to the senior project draughtsman afterwards and saying, 'I think we have found a whole new construction technology which will know no bounds.'

However, it took another two months of intensive design innovation to come up with a system that could be patented as a construction system for general use, long after the A19 Tees Report had been submitted.

I can remember lying awake night after night trying to find the perfect answer. My colleague, the late Roy Templeman, did the same and between us we came up with the answer, later named the advanced composite construction system, ACCS. We needed a system that could be bonded as well as have a sliding assembly and this was the real problem. Now it seems so obvious!

The innovation process had only just begun and it took another six years of effort before the first sections were being produced and erected on the A19 Tees Viaduct. Most of this time was filled with frustrating uncertainty and resistance to change at every turn. We had to produce a completely new design method and manufacturing specification which did not exist for these materials anywhere in the world. We then had to convince clients and their advisors that the system would perform as predicted with no prototyping. The greatest help came from Westland Helicopters at Yeovil who freely gave us their time and experience, having heard of our work through the British Plastics Federation.

The greatest resistance came from within the construction industry itself, which until recently has seen the introduction of new material technologies as a threat. We have been the subject of a number of parliamentary questions over the years, with people even questioning our motives and technical competence to move the new ideas forward.

The A19 enclosure project was completed to time and budget in 1988 and has been a huge success, providing excellent value for money. It is still the largest single use of advanced composite materials in construction anywhere in the world and shows the power of British technology when harnessed from a range of different disciplines (see photograph opposite).

A19 TEES VIADUCT – ENCLOSURE.

Our proven track record in innovation helped Maunsell to win the commission in 1984 to study new crossings of the Severn Estuary. I then had the privilege to play a key role in the project until the completion of the new Bridge Crossing in 1996. My tutor at Imperial College, Dr Tom Wyatt, joined our team and began a working association over 12 years which led to the successful implementation of windshielding on the Second Severn Crossing. It was vitally important that the new crossing should be available for public use in all weathers, and a more expen-

81

sive tunnel would have been preferred to a bridge unless traffic could be windshielded with confidence. This story has even more twists and turns than the earlier ones, particularly when we were involved in a fierce battle in Select Committee over the use of windshielding on the Dartford (Queen Elizabeth) Bridge. Acting for Kent and Essex County Councils, we were proposing the use of the windshielding concept already developed for te Severn Crossing. We spent 19 days in Parliament locked in vicious argument and this had a profound influence on me. It turned me into a 'street fighter' for innovation and change. We actually won the argument in Committee, which ruled windshielding should be included, but lost the war because the large Government majority at the time overturned the ruling on the floor of the House. As I sit writing this chapter, the newspaper headlines in front of me read, 'For the second time in three days, Dartford Bridge was closed to traffic for several hours because of severe storms, while Second Severn Crossing stayed open'.

Successful innovation in engineering improves the quality of life for everyone, combining safety and convenience. Every time storms sweep up the Bristol Channel I get a real sense of achievement to hear that all traffic is crossing the new bridge safely. Wales is now always fully connected to England (see below).

M4 SECOND SEVERN CROSSING.

Another of the many innovative features we brought to this project was detailed attention to the impact of construction on the environment. One of the joys of working on site is to live closer to nature and to have the opportunity to observe animals and birds through different seasons. As long as care is taken to control pollution of soil, water and air, nature can thrive around major bridge sites. For example, I have seen herons swooping and diving around several bridge sites, apparently enjoying the new air currents and deep shadows (as in the photograph below). When we were struggling with environmentalists on the Northumberland Straits Crossing Project in Canada, who were claiming that extra bridge piers would destroy the lobster fishing, we were approached in our hotel rooms by local fishermen wanting to secure the rights to fish at the piers because rock armour placed around them would provide excellent breeding conditions. Engineers are always at the cutting edge between people and nature. Many of the great innovations in the future will be connected with the successful management of this interface, improving quality of life for people and protecting the environment.

The pinnacle of successful innovation so far in my career began in a hotel lobby in London's Victoria in September 1991. A good friend from Freeman Fox days, Bill Harvey, who was lecturing at Dundee University, rang and asked to

FRIARTON BRIDGE, PERTH.

meet me to discuss a possible project. He had read about a lecture Brian Richmond and I had given the year before, saying that we could now design and build a major footbridge using advanced composite materials and were looking for a site. Bill had been approached by the managers of Aberfeldy Golf Club asking for help from the University to build a footbridge across the River Tay to enable them to extend their course from 9 to 18 holes. They had little money and the bridge needed to span over 60 metres because of ferocious winter floods coming down the mountain valley from Loch Tay. Sitting in the rather frayed Edwardian hotel we decided there and then to go forward. I approached our friends in industry and invited them to a meeting at the Golf Club in November. It was a typical freezing Scottish day with thick cloud cloaking the surrounding forested hills and the river in full spate. We sat in the snug club bar, with windows steamed up, and said Maunsell and the third year civil engineering university students would design and build the world's first major all-composite bridge by the following October, if industry would provide financial and technical help. They all agreed, although I do not think many of them thought it would actually happen. And so began a voyage into the unknown. We planned to build a major bridge, 120 metres long, using materials weighing a fraction of any used before, using adhesives for all connections, using a team of 10 students with no civil engineering plant and employing a construction method for launching a cable stayed bridge deck never used previously.

In the Spring we were on site with a colour chart matching the polymer colour of the composite structure to the colour of the local tree branch tips and in March drawings were issued for planning approval. This was true innovation where the materials, structural form, aesthetics and construction method were all new. We now knew how Darby must have felt as Ironbridge was conceived and built. We encountered huge problems and overcame them all by personal commitment and endeavour by a superb team. We built the bridge ourselves, piece by piece, and planned the day when we would launch the deck over the main span, even inviting BBC's *Tomorrow's World* cameras to be there. We needed a calm warm August day — we woke at 4 a.m. to find a 50 mph easterly gale blowing with torrential rain. We struggled out in our waterproofs and worked non-stop until 12 noon when we planned the launch. However we needed calm winds. The local Met. office said a deep depression was crossing right over us during the day and the eye of the storm would pass over Aberfeldy, giving calm between 1 p.m. and 4 p.m. when a westerly gale would set in. I took the decision to launch in this window, knowing that if anything went wrong we would be in serious trouble.

Sir Hubert Shirley Smith said at the Enquiry into the collapse of Milford Haven Bridge in 1970, 'Although factors of safety universally considered adequate are included, unforeseen hazards may be encountered in pioneer work that is pushed towards the limits of engineering knowledge.' We took the view that you could not leave anything to chance when the eyes of the world were on you. You had to try to think of everything and we were confident we had modelled the launch well and it would work.

ABERFELDY BRIDGE.

At 2 p.m. we were ready at last. There was an eerie calm, just the drumming of the rain on the umbrellas of the considerable local crowd and the rather damp and grumpy television crew (see photograph opposite). Bill Harvey standing on the end of the bridge above us raised his hand to signal the start of the pull from the opposite bank. The river was swelling with every minute, threatening to flood the site. There was an electric atmosphere and looking along the line of onlookers some of their hair was standing on end! As the bridge jerked forward there was a flash and huge bang as a thunderbolt burst overhead amplified by the surrounding hills. It was a Wagnerian moment of true drama. In that instant human engineering spirit was somehow joined with nature and the endeavour shown by the team still shines there through the completed bridge. It is a beautiful structure and has become world famous through television (see photograph over).

A year after the opening, I visited the site to show a group of Japanese engineers the bridge. At one end, sitting in the warm autumn sunshine, was a couple, the woman playing a guitar. I said hello and asked where they were from. They were from Memphis and decided to come and see the bridge they had heard so much about. Then coming across was a thin, exhausted-looking cyclist from Milton Keynes who was visiting the bridge as part of a cycling tour. We all had a laugh and an impromptu karaoke with the Japanese engineers. Innovation is like that.

ABERFELDY BRIDGE.

Peter R Head

OBE, FREng, FICE, FIStrucE, FIHT

Peter Head graduated from Imperial College in 1969 with a first class honours degree and spent the early years of his career working on the design and construction of major steel bridges. By the time he joined Maunsell in 1980 he was well-known in the construction industry as a steel bridge expert, lecturing widely on the subject.

Peter's innovative design side flowered in the Maunsell culture and he was soon leading developments in the use of reinforced polymer composites, in parallel with the technical input to and subsequent management of major bridge design and construction projects such as the Second Severn Crossing, for which he became the Government agent. He also led the team developing designs for the Forth Crossing, Dartford Crossing, and Northumberland Straits Crossing, Canada and was project director for the Kap Shui Mun Bridge in Hong Kong. In 1995 he was the first recipient of the Royal Academy of Engineering's Silver Medal for an outstanding personal contribution to British industry. He was elected a Fellow of the Royal Academy in 1996, was promoted in Maunsell to the position of Chief Executive of the European Group in 1997 and received an OBE in the 1998 New Year's Honours for services to bridge engineering.

Also in 1998, Peter was appointed Laureate of the 1998 IABSE Award of Merit. He is only the second British engineer to receive this award in over 20 years, the first being Oleg Kerensky.

Teaching yourself to think

Peter Hewlett

Introduction

The invitation to put together this chapter could not be resisted since it represented an opportunity to write in a personal and perhaps subjective and biased way my opinions and impressions about the role of thinking in engineering, civil engineering in particular, and indeed the process of thought itself. As a non-engineer but a scientist, I can exercise a degree of naivety that is excusable in commenting on the role of thought within the activities of civil engineering and building.

Engineering and science demand a two-fold allegiance, on the one hand respect for conventions of the day and on the other, new principles and novel solutions. Reconciling these apparent contradictions requires thought and therefore the activity of thinking and its result are worthy of consideration that is not unique to civil engineering. Since that represents a familiar area for me, and a pertinent area of interest for this chapter, I will keep that aspect in focus.

The activities of civil engineering and building are infinitely variable, calling upon a range of technical, organizational and administrative skills. They are creative activities that leave a permanent mark and provide useful private and public services. To be part of such activities can be rewarding, stimulating, demanding and frustrating. However, it is never boring.

Notwithstanding my father being in building and making it very clear he did not want me to follow in his footsteps, I look back over 34 years of involvement as a qualified chemist, having worked with and for engineers within the construction and building sectors – an unexpected but fulfilling outcome.

Having left university with two degrees after six years, it was perhaps understandable that I felt a period of relief was justified. I obtained a job as a research/development chemist with the then Cementation Company Limited* (readily obtained in 1963, so different now!). It took me all of three weeks to realize that engineers at the sharp end of the activity expected miracles, did not understand

---

* civil and specialist engineers

the materials they were using in abundance, gave you responsibility and assumed your enthusiasm matched theirs. It was quite an emotional and technical mix and I wanted none of it. The outcome nevertheless was that I stayed for almost 25 years, since the combination proved to be a heady one with recognition and advancement being swift.

My present position as Director of the BBA* has maintained the same level of motivation and, since both involvements relate to civil and specialist engineering and building, I have concluded that stimulation is a characteristic of these activities and not simply a characteristic of me! If that is true, I hope some of that enthusiasm will be passed on to you.

By not being an engineer and yet trying to introduce new technology to engineers, I had the opportunity of trying to communicate between recognized but very different disciplines and to work across conventional boundaries. This made me talk about my subject, that had now become more materials science than chemistry, not only to the users of the newer technologies but, perhaps more important, to myself.

Convincing yourself is the first step to convincing others and to talk with them rather than at them. So after seven years of secondary schooling and six years at university, I had to start thinking for myself rather than learning from others. I will be always indebted to the civil and specialist engineering environment for producing this result. It is one of the characteristics of this industry that young people can respond to. If you are willing to take part you will be welcomed and stretched. It is commonplace for engineers at all levels to talk about their work with, on occasions, youthful enthusiasm. It is creative and they are conscious of their contributions. Such an environment can be very motivating for a young and relatively inexperienced person. It was so for me and still is.

I referred earlier to thinking in both general and focused terms; let us take the matter further.

Working at the interface between research and its practical application highlighted the problems of presenting information in a way that was clear and true and yet would also convince. This was a gradual move to self-awareness that was quite revealing. Much of what I had learned was protectively cocooned in the language of specialists such as chemists, physicists and even medics, and yet practical development and transfer of this information required representation and orientating in such a way that convinced others and at the same time myself. It is very easy to deceive a class or group of people into accepting that a subject is complicated and that will surely alienate most of your audience. However, if you are sure of your subject and have honestly pursued understanding it, then simplifying the language, establishing the core issues and removing the fringe clutter are quite easy. To do so requires a demand on oneself to re-look and re-think the issues. Teaching yourself to think is at the heart of this process. Can it be taught?

---

*British Board of Agrément (certifying authority for innovative construction and building products).

What are the essential aspects of doing so? I do not mean hopeful daydreaming or serendipity but a logical collecting together of information in a disciplined and questing way. If you do it, it is usually welcomed but can also be regarded as threatening to those who feel secure in the conventions of the day. Let us explore the activity of thinking further.

Convention and conformity

If we are to encourage people to think for themselves and make their opinions known, we have to be prepared for the popular reaction. It will often be negative and sometimes hostile. The maintenance of conventions is what is familiar and underpins a feeling of security. Again, we have a dilemma since progress, particularly radical progress, usually means breaking conventions. For instance in the field of astronomy we have the familiar (secure?) view that life originated on planet Earth. Many social and religious norms depend on this view. However, there is growing evidence that life on earth is not a unique phenomenon and may have originated from cosmic dust and encounters with migratory comets. Scientific and lay responses to this alternative explanation of life have been mixed. This is a radical example, but serves to warn you that free thinkers and breakers of conventions can have a rough ride. Even so, and at a more modest level, you should endeavour to present and explain new ways of doing things and go to some length to reassure and include others in the presentation process. Intellectual arrogance is not an endearing characteristic and if your concern is to make progress then you will want to win the argument and see your ideas taken up. Self-satisfied detachment will not achieve that objective.

I have always thought it strange and somewhat inconsistent that to improve one's performance at almost any activity one has to work at it time and time again, and yet when it comes to intellectual performance there seems to be a resignation to whatever the genetic lottery has given you. If you practise intellectual activity you run the risk of being labelled a swot and stepping out of line with current norms. But children and adults are encouraged to practise music, train for athletics, take singing lessons, learn the skills of chess or even conjuring.

Therefore we should dismiss the notion that we cannot improve how we think or mobilize our thoughts simply because some of us take longer to do so. Nor should those who are modest, polite and deferential rather than aggressively assertive be considered less capable. Do not confuse saying a lot with having something worthwhile to say.

Thinking is the process of considering and judging information in order to draw balanced and logical conclusions. When one is young the presentation of information will either stimulate or bore. Both responses are important and govern whether we take the matter further or ignore it.

Unless the information is considered relevant in some way, stimulation may only be achieved through threat, e.g. learning one's times tables. Relevance is a key to interest which in turn is a key to retention of the information. How do we engender relevance? Much 'trendy' teaching has been preoccupied with confusing amusement and play with relevance. There is a risk of trivializing the information by adopting this approach. An alternative is to try and think as a child might think and relate the information to daily events that can be recognized. Project work dealing with local matters can assist. This is the basis of the enquiry approach to problem solving.

Associating relevance with learning is important and particularly so when relating research results to engineering benefit. Working at a science/engineering interface demands this approach and those who can acquire the technique or who may have it naturally will succeed.

Distinguishing between learning and understanding is also a problem. Learning can be similar to training: you are taught what to do and when to do it because there are recognized signals. To participate does not mean that you comprehend. All is well until the signals are not quite so recognizable and you have to make a balanced decision on the information to hand. To do that you have to understand. Understanding resulting from thinking is not the same as knowing. However the two are linked because you contribute better if you have some facts to build upon. Having some knowledge of a subject improves the quality and speed of the thought or deduction processes. So what is the link between obtaining facts and being able to do and the process of thought? I would suggest it is experience, and how we are affected in attitude and response by the experiences that we have. How often have you heard someone lacking in formal education describe themselves as graduating from the university of 'hard knocks', or the university of life, implying that such a routing is more practical, down-to-earth and therefore more useful than book learning. Instead of these two views being seen as alternatives and competing, they should be seen as complementary.

The role of experience

Very young children are exposed to events which seem to leave a deep imprint, that becomes layered over by other events and impressions until these events are buried deep in our psychological make-up and yet continue to affect how we react to new situations in subtle and even denied ways. Our past experiences seem to manipulate our current attitudes and responses, a 'skeleton in the cupboard'.

In order to encourage enthusiasm for learning (acquisition of facts) and a willingness to think (to cross-question and reason from the facts), early experiences should be of themselves encouraging and supportive. Of necessity, educational or academic progress has to be structured and monitored, and one of the consequences is that judgements can sometimes be sufficient to discourage further effort, and the process of learning becomes a chore instead of a stimulation. Failure rather than attainment becomes the label, and from that point on indulging in intellectual pursuits becomes pointless, simply a recipe for further humiliation by academic peers. The consequences may be to change one's peers to less critical ones, with the added prospect of social detachment also – a not very enlightening sequence of events.

Early stimulation to learning and the ability to think that flows from it are important. Being aware of one's ability to think can generate its own momentum in the same way as improving one's time to run 1000 metres can spur one on to greater effort and the emergence of pride and a sense of achievement.

Now a dilemma arises, since achievement is often compared with that of others and this tends to create competition and the risk of failure. We have seen over the last 20–30 years a wish to remove any sense of failure or competition in the educational process, and yet we encourage it in athletics, music and even quiz games. It is possible to compete against oneself, as many amateur athletes do, and in this way maintain an enthusiasm for involvement. To think requires a certain maturity and self-awareness usually absent until the teenage years. Many subjects including mathematics, physics, chemistry, materials science and engineering are demanding, both personally and collectively: they require hard work and application in order to mobilize the facts.

Good teachers are well aware that the core subject matter requires application and hard work. Some teachers are well endowed with clear thinking and can pass it on to others. In this respect, however, some are less able. Early discouraging experiences in learning can cause indifference towards learning later. These early experiences provide a blueprint or framework for later life and can be built upon layer by layer as you gain experience.

It is important to think around a subject as well as along it. One of my first projects for the Cementation Company concerned a group of resinous organic materials called epoxide resins. I had never heard of such materials but was charged nevertheless with developing a soil/rock stabilizing mixture that could be used in wet and dry conditions above and below ground. It became clear that this problem involved chemical kinetics, solution/dissolution mechanisms, adhesion, spreading and wetting, chemical selection and associated health risks. It was not practical to divide the project into these various aspects and I had to start learning and thinking about other matters than those I had been doing for a PhD. What I did not realize at the time was that the process of doing a PhD had already shown me how to search out information, assemble it in some sort of order and look for relevant interconnections. It was also necessary to talk to other experts and subject my suggestions and thinking to their scrutiny. This was revealing and sometimes embarrassing since one's emerging confidence was readily damaged. It was not long before I could compete and found the process of acquiring new viewpoints really very stimulating. It was a style of working that remains with me to this day and it alarms me on occasions when young professionals are not prepared to change or chance their thinking and want to doggedly pursue a deepening but very narrow furrow.

Because the activities of construction and building are so diverse, many skills are brought into play. Whilst core disciplines tend increasingly to demarcate, no matter how you enter a career stream, e.g. design engineer, estimator, geotechnician, mining engineer, etc., it is very likely that people will move around and broaden their experience. The demarcations then become less obvious. Often you do not know what you can do until given the opportunity.

This is a double-edged opportunity since a move into a less familiar area of work will expose you to possible failure as well as reward. As people mature they are better able to make judgements and to view decision-making from a broader perspective. The bringing of these aspects together still requires thought and thoughtfulness. You have to use information, perhaps even generate it to begin with, in order to meet a given objective. Some people have this skill in an intuitive way, others have to learn it and some never acquire it. Indeed the 'physical manager,' whilst considering himself resolute and decisive, would dismiss the thought of consultation and consensus as weak-kneed and namby-pamby. I would encourage good communication and a willingness to persuade coupled with a dogged pursuit of the objective. It would not be unusual to discover you understand a matter better as a result of having to explain it many times to doubters or by imagining you have to teach or explain it to novices. One has to strike a balance between enthusiasm for a result and the need to carry opinion with you. Construction and civil engineering activity can offer opportunity to develop these skills and one should not be afraid to branch out.

Your initial discipline can sometimes constrain subsequent personal career development. Technical qualifications can militate against developing laterally. An engineer or scientist when newly qualified will wish to practise and apply the skills acquired and that is how it should be. But for how long? In my own case I had a view to work at materials research within specialist construction for at least 10 years after obtaining my PhD at approximately 24 years of age.

Within six years I had the opportunity to move into more commercial technical and general management albeit within the same group of companies served by the research department. This move carried with it materials inducement lacking in a research activity. I turned this opportunity down on the basis of wanting to maximize my research contribution before broadening and generalizing. It was nevertheless a tonic to be thought of in this way by my employers and I hoped that later on other opportunities might arise. They did, ultimately resulting in six directorships. You might argue that this was the fortunate result of being in the right place at the right time. However, I know no other activity than construction and building that could have offered such variety and opportunity. You should take encouragement from that.

There is an alternative view whereby many individuals stay with a particular discipline for the span of their career. Some obtain just as much fulfilment without diversifying and pursue related interests in trade associations, professional institute activity and so on. If this routing is from choice, then extending into more general management might well be thought of as retrograde. You have to decide when change and opportunity present themselves and if they are right for you. Be prepared to branch out. Such decisions require thought and some courage.

It is fair to say the material rewards for extending into more general senior management are usually greater but so are the demands, responsibilities and risks. However, satisfaction and career fulfilment are subjective and can take a variety of forms. Whatever one decides, it should be done as well as you can and with thought. You can never avoid the need for thought both for yourself, your dependants, your superiors and indeed those for whom you are responsible.

It is always difficult to know just how relevant your own personal experiences are when passing them on to another person. In this regard the response will always be subjective. Even so, having spent my entire working career employed by industry, I have always found it constructive and necessary to maintain very close contact with academe. The activity of lecturing, teaching and advocating a given subject has always had a strong appeal. As a postgraduate studying for a PhD I undertook external lecturing (mathematics and physics, not chemistry!). Leading these two existences at the academic/industrial interface has maintained my sense of curiosity.

This is helpful in achieving a balance between practical application of knowledge and new ideas and their generation to begin with. A very necessary form of masochism. I have been fortunate indeed in having close contact with a number of universities, and Dundee University, Department of Civil Engineering, in particular. Again this brings into play interfacing skills; in addition, contact with students and academic staff keeps one 'on the ball'. The opportunity to lecture is also important to me and adds to the involvement in communication – a necessary honing stone for ongoing thought process development.

The opportunity to lead this dual existence may not appeal to everyone, but it is pertinent that the opportunity was offered to a non-engineer by engineers and I think that reflects the multi-faceted nature of engineering as opposed to other disciplines.

Concluding comments

I believe one can be taught to think and learn from experience and the acquisition of knowledge. An informed response is so much better than an intuitive one.

Civil engineering and the related activity of building is a creative process requiring the assembling of ideas and options that have to transpose into hardware. As well as being creative it is multi-disciplined with opportunities for all types of technical skills.

Many engineering objectives can be met by the application or adaptation of existing knowledge, but you have to have that knowledge or the inquisitive skills to go and search for it. Such interfacing of knowledge and practice is one of the hallmarks of engineering. It is never boring and may well become a lifetime's quest.

Think about it.

Professor Peter C. Hewlett

BSc, CChem, PhD, DDL, FRSC, FIM, FInstConcTech, FCS

Professor Peter Hewlett is a chemist by qualification (graduated in 1960) and obtained his PhD in 1964 in the field of physical/organic chemistry and specifically transitional metal complexes having potential biological activity. However, his career has been in materials science research and for the last 10 years with performance-based assessments and certification of innovative construction and building products. Specialisms include adhesion, hydrogel polymer systems, cement and concrete as well as high-performance synthetic resins such as epoxides, polyesters and polyacrylates.

For almost 25 years he worked with Cementation Research Ltd (from 1963), becoming its Research Director in 1975. He held several concurrent directorships with subsidiary companies covering speciality chemicals and ground engineering processes in what became the Trafalgar House Group. Since 1988 he has been Director of the British Board of Agrément. Since his postgraduate period he has always worked with universities at both a research and teaching level. He has been the Visiting Industrial Professor to the Department of Civil Engineering at Dundee University since 1986 and was awarded a Doctor of Laws (*honoris causa*) in 1993 for his work, particularly in the field of concrete durability and surface properties.

A previous Chairman and/or President of the Cement Admixtures Association, Research and Development Society (UK), Concrete Society and European Union of Agrément. He is currently President Designate (second term) of the Concrete Society and President of the European Organisation of Technical Approvals. A member of many committees (FSRC, EPSRC, CIRIA, Concrete Society, etc.), in particular Chair of the editorial board of *Magazine of Concrete Research* (since 1987). He has lectured and published widely and internationally.

95

The impact of reality: design by test instead of text

James Sutherland

The development of Laingspan structural system as a career-forming experience

The event which changed my ideas on engineering more than anything else was an introduction to experiment rather than texts as the basis for structural design. This totally altered my attitude to precedent, acknowledged authorities, textbooks and, above all, codes of practice, although at the time codes were less dominant than they are today.

I was then in my mid-thirties and had spent three years in the Navy and ten with Sir William Halcrow & Partners. My time in the Navy, starting towards the end of the War, was more notable as sponsored travel than for naval action but did include a revealing two or three months 'occupying' Japan immediately after the surrender, seeing Yokohama, Tokyo, Kagoshima and Nagasaki. If observing devastation of both solid and flimsy construction was good training for future civil engineering, I had plenty of that. My most abiding image is of a pile of bottles at Nagasaki fused together by heat into a form like an abstract sculpture but with its origins still just discernible. As we walked around at Nagasaki we had no inkling of the health problems. As far as we knew it had just been a very big bomb.

Real civil engineering started with Halcrows on hydro-electricity schemes and steam power stations. Following a dispiriting first week or two proof-reading contract documents – a useful initiation although it did not seem so at the time – the work there became increasingly interesting. The firm was very friendly and quite small in those days. I remember conker matches in the autumn against Kennedy & Donkin from the floor above, nominally in our lunch breaks! Charles Haswell, who later founded Charles Haswell & Partners, was my immediate superior and mentor and remained a friend until his death in 1989. With Halcrows I had an equal time on sites and in the office, finally as leader of quite a large design team in London. No complaints here, but ten years in one job seemed a long time and I had a yearning to catch up with the newer materials and techniques which were becoming more widely recognized: laminated timber, welded steel, composite constructions and above all prestressed concrete, hardly Halcrow's forte in those days.

Several strings were pulling me towards the new developments but one conversation during my time working on site stands out. It must have been around 1950. I was standing with the resident engineer, my immediate boss, on the massive reinforcement cage of a pile cap when he turned to me and said: 'Sutherland, I wish I understood what these bars really do.'

This casual remark was a revelation. He was no fool and was well-respected by the contractor and by his own staff, but it seemed that he had missed out on reinforced concrete in his education and not picked it up later. Naively and tactlessly I entered upon an explanation of the elementary theory. He was very nice about it but then he was a very nice man. I say 'was' because he was then in his early seventies and that was fifty years ago. Memories of this talk kept coming back. Was I in the 1950s in the same position in relation to prestressed concrete as my resident engineer had been to reinforced concrete when it was in its infancy at the turn of the century? I simply had to learn about prestressing before it was too late.

In 1955 Charles Haswell introduced me to Alan Harris (now Sir Alan) who agreed to take me on for six months knowing that I had another link-up in mind. Alan had just set up as a consulting engineer after some time as Director of the Prestressed Concrete Company. He was working on his own in his flat with his wife Mathé as secretary and one young Australian, Raleigh Robinson, on a short-term industrial exchange scholarship.

The change from my work at Halcrows was startling. Here was not only the new technology but innovative thinking and a wholly new attitude to design. For instance we went to the Farnborough Air Show, not for the aeroplanes which were incidental, but to see if there were any new materials on the suppliers' trade stands which might be useful in the future for our types of structures. Alan's practice grew. I stayed beyond my six months. Two years later I was a partner with him and his brother John. Thirty years later I retired from Harris & Sutherland, then quite a substantial firm.

By far the most 'mind-forming' project for me was one of the first, the design of the Laingspan structural system for school buildings. This was very largely developed on the basis of physical testing. To appreciate why this was so formative it is necessary to understand its administration, which was unusual, and at least some of the details of the structure.

The design team consisted of the Development Group of what was then the Ministry of Education (MOE), John Laing Research & Development and Alan (A.J.) Harris as consultant on the structure.

The MOE Development Group was staffed by socially-minded architects and quantity surveyors, together aiming, by example, to set high standards of functional efficiency and low cost for local education authorities to follow and on which cost limits could be set. Aesthetics in the conventional sense hardly came into the thinking. I can scarcely remember seeing an elevation of a building there. All was

97

done on plan. Saving of space was the goal. Nevertheless the buildings were far from ugly but comely, cheerful and well thought out within the functional tradition of the day.

John Laing Research & Development was a small but exceptionally creative research unit set up to provide a service to the contractor. Although the equipment was far less lavish than that in the larger university laboratories, the staff had a great capacity for improvization and worked to high standards. Architects and engineers from this group were also very much part of the Laingspan design team, especially on items like cladding.

Alan Harris was engaged to provide an economical structure, systematized and prefabricated for speed of design and building, flexible in planning terms, light in weight to save on foundations and above all using the minimum of steel, which was still in short supply. It was a curious feature of the time that structural steel sections and reinforcing bars were on licence but prestressing wire was freely available. Hence the attraction of concrete and the great appeal of prestressing.

It was Laingspan (*see* Figures 1 and 2, courtesy of John Laing) which provided my entry into this milieu and the development of Laingspan which proved such a

FIGURE 1 *A Laingspan structure during erection.*

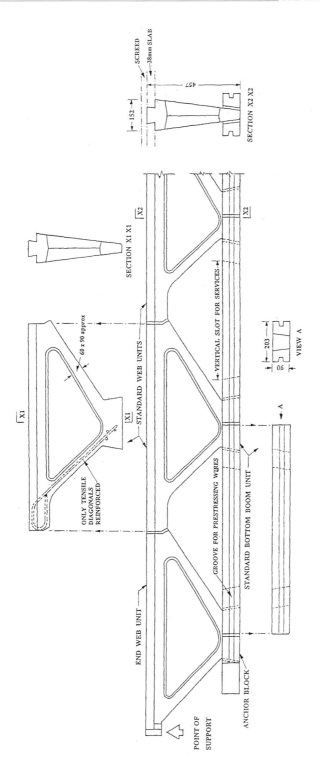

FIGURE 2 *Precast concrete floor or roof trusses for the Laingspan system in its earliest form. Each unit was one module (1.02 m) long, providing spans of up to 10.2 m for floors and 14.3 m for roofs.*

liberal engineering education. When I joined him, Alan had fixed on the basic structural form for Laingspan and I am claiming no credit for this. However I was increasingly drawn into the testing of it, the adjustment of its details and its later development. This is where my 'education' took place and I am sure that I learned far more, even in the first year on Laingspan, than I could have in a decade in the lecture room. Through these tests we were learning truths from scratch, whereas in university lectures the 'meat' is almost always pre-cooked, flavoured and re-heated, sometimes well and sometimes less so.

The first lesson – not just from Laingspan but primarily so – was that there is no mystery about the theory of prestressed concrete. It is just a case of applying forces to the concrete through tendons to counteract the forces due to gravity or other loads. However it is an active approach rather than the more usual passive one of providing enough strength. I was lucky in that at the time of my introduction to prestressing there was no relevant code and thus none of today's plethora of interlocking clauses to confuse this simple concept. It was over the details that I found you needed to be a little cunning. The second lesson, soon after, was that all structural behaviour could best be visualized in terms of paths of forces on and within the material. The formulae could come later, when and if necessary.

A further lesson was that concrete sections carrying large loads could be very small, especially with prestressing, and certainly far smaller than I had been used to. We tend to forget how little of the concrete in reinforced concrete actually contributes to the structural action, a large part being associated with bond to the bars and perhaps even more with protective cover to these. Eliminate the tensile forces in the concrete by compressing it and the need for reinforcement vanishes. Minimize the area of concrete and you minimize the prestress needed and thus the cost of the tendons. High concrete strengths and structurally efficient cross-sections become especially desirable. Given essentially direct forces, as in trusses rather than beams, one only needs to prestress those members otherwise in tension. All this, which may seem very simple today, was a revelation in the 1950s. Conventionally with reinforced concrete we put some nominal reinforcement in every section, whether needed or not. If left out, most often someone told us to put it in. It was Laingspan more than anything else which provided me with the antidote to this attitude.

The Laingspan System

Put at its simplest, Laingspan consisted of extremely slender precast concrete columns, together with floor and roof beams, which were really trusses assembled on the site from small concrete sections stressed together only along their bottom edges and with no reinforcement except in the tensile diagonals (too short for prestressing). Precast concrete slabs only 38 mm thick spanned between the trusses for the floors and compressed straw board named Stramit for the roofs. The basic module was 3′4″ (1.02 m), at which the trusses were spaced, maximum floor spans being 10 modules and roof spans 14 modules with a possible building height of four storeys. Apart from foundations the only 'wet' materials used on the site were in the joints between units and in the thin floor screeds. Although only 1'6" (457 mm) deep, the trusses provided quite generous spaces for pipes and cables to pass through them.

Figure 2 shows the principle of the truss with some typical dimensions of the concrete sections, while the photograph shows several of these trusses during erection. If built in steel the dimensions of the sections would not have been remarkable, but in concrete they seemed unbelievably small, or at least they did to me compared with the massive civil engineering scale I had been used to, and the whole thinking was unconventional by any standards. Clearly all this could not be justified on the drawing-board alone.

We drew concrete units, they were cast and set up in Laing's laboratory where later all concerned watched them being loaded to destruction. This was an immensely exciting process. The first creak heard increased alertness. One wondered when and where the first crack would appear and what it would tell us. Then how would the assembly fail and at what load? Here we were aiming for a factor of safety averaging two but the design was too near the limit for immediate adequacy to be expected in all these trials. Once a test showed a weakness we adjusted the design to eliminate it, or so we hoped, despatched revised drawings and within days were called to see the revised version set up for a further test. Anticipation tended to be even more acute than on the first occasion. Perhaps all was well. Perhaps some further small adjustment was called for. Similarly the columns (pre-tensioned but unreinforced) were tested and all aspects of the connections and of composite action, such as how the bonded Stramit transferred wind loads to the minimal number of vertically-braced bays.

No one could doubt that this was a highly educative process. We were seeing how the materials really behave and not how the textbooks told us they should. Further, there was an immediacy. This was no academic exercise, but justification of a substantial investment with a tight deadline for completion.

Having proved the 'bones' of the system one might think that all was over. Not so. Once you start to use any building system the unresolved details start to emerge and each has to be cleared for the whole system and not just on a job by job basis. We produced at least half a dozen 'standard' precast staircases in detailing the first project.

One feature of system building which tends to be overlooked at first is the urge to go on developing it. This seems universal. Not only do individual architects using the system ask for 'just one small change' but the promoters and designers themselves see potential benefits from improvements and manufacturers see simplifications in construction. This is where I became increasingly involved. Again it was highly educative. There was a great cooperative feeling in the team that we were creating something worthwhile. Ideas emerged, were discussed and tried without it always being clear who had first thought of them. In one instance I was asked to develop a deeper roof truss spanning 30% more than the limit originally planned. This was initially for one school but seen as part of the system. It was used once but never again.

There was a suggestion that we could save money by halving the number of trusses in the roofs. This led to a very simple roof panel spanning two modules and made of a material akin to chipboard with minimal stiffening added. It could not be justified on a code of practice basis but was developed and proved by test. Someone then suggested that the whole roof could be of timber and I spent some time designing a glued timber truss of exactly the same shallow profile as the prestressed concrete one and designed to act compositely with a plywood deck. Again this needed to be tested. It proved to be strong enough and gratifyingly stiff but failed in a more brittle manner than that usually associated with timber. I think it was cost which actually killed the idea, but the insight into the behaviour of highly-stressed timber was invaluable. These are just examples.

Trade practices were evolving fast by the 1960s. Plant for lifting and transporting bulky units was increasingly available. With Laingspan this evolution was reflected in the change from floor and roof trusses, made up on site from units which could be lifted by hand, to single full-length pre-tensioned ones of the same profile delivered complete by road from the precast yard and just lifted straight into place.

Some fifty structures, mostly schools, were built in the Laingspan system but its success commercially is debatable. It was competing with Intergrid, an earlier MOE school system with a concrete structure originated by Alan Harris and later with rivals in steel like CLASP which were being given considerable official backing in the 1960s. With plentiful steel (easy to fix to!), very light concrete lost much of its attraction. Laingspan faded out quietly as gradually did the whole zest for system building. All concerned – designers and builders – were becoming increasingly involved in other fields. However in the present limited context it is the educative value of the development of Laingspan which is relevant (at least for me) rather than its commercial rating.

Conclusion

Laingspan was not my only new commitment, but its development led to new beliefs and a new way of thinking about how materials and structures behave. The idea of a single *right* way of approaching any design could no longer be contemplated, analytical methods laid down in textbooks and codes being only convenient models by which routine structures can be made to work. Clearly the 'answers' given by such analyses bear little relation to the way in which structures actually behave. Once built, a bridge or building will decide for itself how it will behave, often far better than conventional models indicate. Further, the structure may well change its 'mind' about its action as time passes and conditions change. Thus, in any design or analysis I believe that we should recognize the limitations of our methods and look for means of extending these limits. Full reality may not be achievable but the search for it can be rewarding. These beliefs have remained with me throughout my career. They may have developed and been clarified but initially they stem more from experience on the Laingspan system than anything else. I recommend such thoughts for consideration by others, particularly those either appraising existing structures or trying to help in the resolution of legal disputes.

At one stage I wondered whether my first ten years of heavy civil engineering had been wasted, but later I decided that they had not. They provided a sound background from which to develop and also something to rebel against. Later, when tackling civil engineering on my own, for roads and bridges and major mixed projects such as the new universities of Essex and Bath, with their lakes, level changes and extensive earthworks as well as buildings, this early experience was reassuring. However, the lessons of Laingspan still applied, although the scale was vastly greater.

James (R. J. M.) Sutherland

BA, FEng, FICE, FIStructE

A civil and structural engineer and partner in Harris & Sutherland from 1956 to 1987 James Sutherland is now consultant to the firm. Essentially a designer, rather than an administrator, he has been responsible for a broad variety of projects ranging from urban roads, bridges and large mixed developments to the conservation of small historic buildings. Especially interested in research into the better use of existing materials, the appearance of all types of structure and particularly in engineering history, now a dominant preoccupation.

He has served on a number of design and conservation committees for the British Standards Institution, English Heritage and professional institutions. He was a member of the Royal Fine Art Commission from 1986 to 1996, President of the Newcomen Society for the Study of the History of Engineering and Technology in 1987–89 and Vice-President of the Institution of Structural Engineers 1980–82 as well as Convenor of its History Study Group from its foundation in 1973 to 1994. He has written and lectured widely on technical and historical subjects

Time and change and structural design

George Somerville

Introduction

As a schoolboy living on a farm in a remote part of Scotland, I was fascinated by bridges. I could not explain it then – even less now, after more than forty years – but it took me to Glasgow University in the 1950s to study civil engineering, followed by a highly enjoyable career where bridges featured strongly.

Glasgow University had a four-year thin sandwich course. Each six-month summer period was for job experience, and two of mine were spent on bridge survey work. This covered relatively new structures in steel and concrete, but also bridges that had been there for centuries, built in stone or masonry. With the benefit of hindsight this gave a feeling for function and time. Neither of these loomed large in the parallel studies at university. There, using the latest theories, methods and wisdom, we were taught how to provide strength, stiffness and stability, at lowest first cost, while being coaxed to introduce a modicum of flair. It was in many ways an abstract intellectual exercise, somehow divorced from the fact that the end result had a function, and was for people to use and enjoy. It never occurred to us that what we designed might have to last for fifty years, one hundred years, or five hundred years. Even if it had, it is doubtful if we would have known what to do; quite simply, function – and its maintenance over time – was insignificant as a design criterion.

Forty years on, all of that has changed; function and the time factor are key elements in design, although we are still not as well tutored as we should be in terms of providing optimum solutions. Forty years is also the average working life of an engineer and, in that time, significant changes can take place in terms of what we are asked to construct, how we do it and the materials we use, and, above all, in the attitudes and needs and expectations of owners and all those who use the output from the construction industry.

Drawing on experience over the busy construction period of the last forty years, the purpose of this contribution is to identify the different types of change, and how these influence structural design both now and in the future. Most of my experience has been with structural concrete and this will be the basis for illustration.

Elements of change

In the construction industry, change is usually slow and incremental; over a period of forty years, however, it can be significant. There are many elements of change, and the approach here is simply to identify the important ones, while giving brief examples of each.

Changes in materials

Materials tend to become stronger. Figure 1 shows the trend of increasing cement strength. Figure 2 paints a similar picture for reinforcement, but showing one added factor: our improved knowledge of behaviour has led to a higher permissible stress in design. The net effect of these trends, linked to more rigorous analysis and design, is that structures are now lighter and more flexible, and possibly more vulnerable to variations in design and construction practices, or less tolerant of

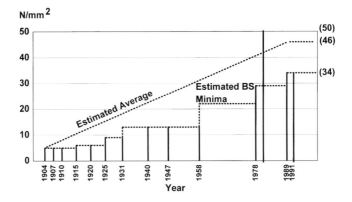

FIGURE 1 *Increase in cement strengths.*

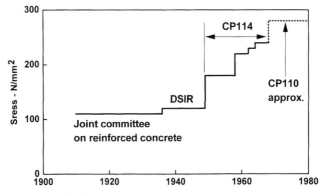

FIGURE 2 *Increase in reinforcement stresses.*

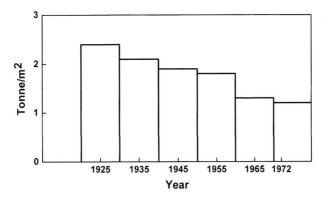

FIGURE 3 *Decrease in building weights (Swedish survey, 1981).*

unconsidered effects in service. This trend is confirmed by Figure 3, giving data from a survey conducted in Sweden in 1981.

These trends with traditional materials are augmented by the introduction of new materials, usually over a significant period of time. Examples are reinforced plastics, the use of fibres, and a host of non-corrodible materials.

Changes in operations and methods

Some examples are given in Table 1. These represent influences both on the nature of the supply chain and on the nature of the final product.

TABLE 1 *Changes in operations and methods (1957–1997).*

	1957	*1997*
Concrete placing	Site placed	Ready-mixed or precast
Section type	Slab; T-beam	Box; flat slab
Prestressing	Very little	Dominant for larger spans (led to new construction methods)
Industry organization	Large number of small companies	Smaller number of large companies (international)

Changes in type of structure

Table 2 gives some examples of what have been more or less continuous trends over that period; there have also been short-term phases, for example the ill-fated industrialized building period of the 1960s.

107

TABLE 2 *Examples of change in type of structure (1957–1997).*

	1957	1997
Energy structures	Hydro-electric; coal-fired	North Sea; nuclear; wind, wave, geothermal
Roads	No motorways	Motorways!
Car parking	Ground level	Multi-storey; underground

In addition, there have been significant changes in the traditional areas of buildings and civil engineering, for example:

Buildings
- More leisure, retail and transport structures; less heavy industrial.
- A shift in the functional requirements inside buildings, due to IT, increased standards, etc.

Civil engineering
- Bridges on primary routes require different treatment, due to the higher risk of obsolescence, and the high cost of remedial or upgrading work, while minimizing disruption to traffic flow.
- The growth of container ports, with road and rail links.

General
- A demand for bigger spans (more usable space in buildings; greater ambition, in terms of what we want to cross, in bridges).

Changes in owners' needs and expectations

Table 3 compares the relative costs of different parts of similar ten-storey buildings, built 25 years apart.

TABLE 3 *Comparative breakdown of construction costs of a ten-storey office block in 1960 and 1985.*

	Percentage of total cost	
	1960	*1985*
Foundations	4.9	2.9
Superstructure	20.3	8.6
Cladding	16.8	28.6
Finishes	24.8	9.2
Services	33.2	50.7

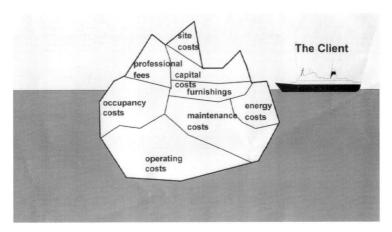

FIGURE 4 *Total life cycle needs: the iceberg analogy.*

The decrease in cost of the basic structures is obvious, coupled with a growing demand for better services and improved appearance.

A second example is contained in Figure 4, which illustrates running costs (compared with initial capital costs), and hence the need to minimize the former. Both these examples are for buildings, but illustrate a general trend: better performance in service, reduced running costs, more focus on maintaining the essential function of the structure in service.

A third example, not illustrated, is the owners' (and public) perception of the performance of our infrastructure in general. There is something of a crisis of confidence in architecture, perhaps triggered by a fascination with the past and disillusionment with the present. There is no consistent style; structures are seen as drab, a situation not helped by durability failures in service. This latter point is conditioned by some of the changes recorded earlier, but is also due to a lack of consideration of the factors which really influence performance in service. We have a history of numerical calculation, given momentum by developments in computer capability. Strength and stresses are important, but feedback tells us that performance in service with time is strongly influenced by environmental effects; these are not always amenable to numerical calculation, but should be a more significant feature in design.

Impact of change and time on structural design

A first indication of this can be obtained by a simple comparison of undergraduate teaching concerns and approach, as given in Table 4. The new (1997) issues have strong time connotations, both during construction and subsequently in service. A major difficulty is that we do not have defined, well-established systems for dealing with this, either in terms of the lifetime performance that we want, or of the processes we need to design for that performance. The time factor introduces

TABLE 4 *Changes in undergraduate teaching concerns (1957–1997).*

1957	Strength Stiffness } via elastic analysis by hand Stability
1997	Strength Stiffness } via limit state design by computer Stability Durability, environment, sustainability, whole-life costing, productivity, performance, cost/benefit, quality; management, maintenance, risk

unknowns and variables, which are not always amenable to traditional numerical approaches. Design has to shift from its analytical base, and be clearer in its intent, while dealing realistically with the factors that truly influence function and performance in service.

That is beginning to happen, and we have clear indications of the key factors involved. Two of the more important are now reviewed.

Environmental loads

Engineers give prominence to self-weight and imposed loads in design. As part of that, wind loading is considered in stability calculations. Temperature effects also come into play, mainly in evaluating temperature gradients in major structures, but also in designing for movement. Wind and temperature, together with the major missing environmental load – water – also have a key role in influencing long-term durability performance, yet are not consciously considered in design.

Different materials are subject to different forms of degradation, with time. The nature and scale of this is now more significant, as more and more man-made aggressive actions are added to those due to nature. An example is chlorides: the use of de-icing salts, containing chlorides, has spread this particular action to a much wider range of structures. We thus have the threat of corrosion on all highway bridges, and not just on maritime structures that are subjected to natural chlorides from sea water.

Bringing these points together illustrates the need to develop a protective design strategy. Buildings, for example, are largely enclosed spaces. An attractive design concept is to protect the load-bearing structure by providing an enveloping cladding system: a design-out approach. Translating that into practical details requires consideration of how wind, temperature and water, together, might breach the protective barrier. Some possible mechanisms are shown in Figure 5; few are currently part of the training and experience of structural engineers.

This type of design-out solution is less easily used for more exposed civil engineering structures. Yet, the basic philosophy is still valid: the recognition of the

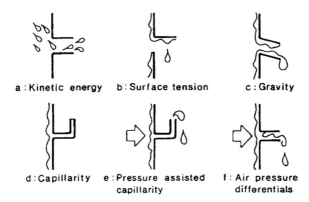

a : Kinetic energy   b : Surface tension   c : Gravity

d : Capillarity   e : Pressure assisted   f : Air pressure
                       capillarity   differentials

FIGURE 5 *Possible mechanisms by which rainwater leaks through joints.*

effects of aggressive environmental actions, and the need to design for these, by some combination of good design, choice of materials and construction quality.

Design life and minimum performance criteria

If we are to develop a design approach for environmental loads as outlined above, we can only do that economically (in whole-life terms) if we have some idea of what we are trying to achieve. Simple as it may sound, this is not easy: different types of structures, the same type of structure with different functions, even the same type of functional structure with different owners, will produce different requirements.

However, if the basic principle is that the technical performance of the structure should not interfere too much with the operation of the building or bridge over a prescribed period, then an attempt can be made to resolve this Catch 22 situation. All structures are made up of numerous elements, and contain a range of services, furniture and fittings. These latter elements have different lives, and have to be replaced or maintained on a regular basis. The trend is to plan for this in a systematic way, at minimum total cost. Table 5 shows such a plan for a building. If the plan is expanded to include the structure, as shown here, then this helps set targets for what is required, in durability terms, while relating that to the management and maintenance of the artefact as a whole. The skill of the designer is then in finding the most economic way of doing that.

TABLE 5 *Basics of a performance profile plan for building*

System	Criticality	Target life before replacement, years						Capital costs*	Cost in use target: x years at y price**	Maintenance requirements
		>5	5–10	10–20	20–40	40–100	>100			
Foundations	A	•••••	•••••	•••••	•••••	•••••	•••	—	—	
Structure	A	•••••	•••••	•••••	•••••	•••••		—	—	
External cladding	B	•••••	•••••	•••••	•••••	•••••		—	—	
Glazing	B	•••••	•••••	•••••	•••			—	—	
Partitions	B	•••••	•••••	•••••	•••••			—	—	
Heating & ventilation	B	•••••	•••••	•••	•••			—	—	
Water, public health service	A/B	•••••	•••••	•••••	•••••			—	—	
Electrical	A	•••••	•••••	•••••	•••••			—	—	
Decorating	C	•••••	•••••					—	—	

Clear specification for each system

*   e.g. 100 units
**   e.g. 1500 units over x years

Concluding remarks

The objective of this contribution was simply to indicate the importance and influence of change and time on structural design; in the space available, only an overall perspective can be given.

It is ironic that, in this age of increasing computing capability, the drive is to wean engineers away from an obsession with analysis and numeracy. The need is to focus on what the structure is for – its function – and widen the brief to achieve that. This is a reversion to the original meaning of the word 'design', to achieve an objective by intent, with a proper consideration of all the factors.

George Somerville, OBE

BSc(Eng), PhD, FREng, FICE, FIStructE, FIHT, FACI, MICT

Born in Dumfries in November 1934, George Somerville is director of engineering at the British Cement Association, formed in 1987 and visiting professor at Kingston University and at Imperial College, London, where he is also active in the Concrete Research and Innovation Centre (CRIC).

After graduating from Glasgow University in 1956, his early experience centred upon bridges and roads, and in 1961 he joined the Cement and Concrete Association in the Structures Department where he became head in 1968, and finally director of Research and Technical Services (1977–87). Research interests included bridges, models, precast concrete, composite construction, durability and reinforcement detailing. His primary interest currently is in durability and design life, and he is editor of a book on the subject, as well as the author of numerous papers and reports. He is lead partner in a major BRITE project with Swedish and Spanish partners on the residual service life of concrete structures. He is also involved in producing European Standards on durability design. He has been involved in code work for many years, particularly BS 8110, BS 5400 and Eurocode 2 and is a co-author of relevant handbooks.

Dr Somerville was elected to Fellowship of the American Concrete Institution (1981), to the Royal Academy of Engineering (1984) and to Membership of the Institution of Concrete Technology (1988). He was the inaugural chairman of the Concrete Bridge Development Group, president of the British Masonry Society in 1990/91, and a member of Council of the Institution of Structural Engineers for nine years, including serving as a vice-president. He was awarded Oscar Faber bronze medals in 1972 and 1986 and the Lewis Kent Award in 1991/92 for services to ISE. He received an OBE in the Queen's Birthday Honours, in June 1997; the Swedish National Concrete Award in October 1997, and was made an Honorary Member of the American Concrete Institute in March 1998.

Diversity and happenstance

Alfred Goldstein

*I*t happened as it did because my sister married an American!

So, aged 22 and some two years after I had started with consulting engineers R. Travers Morgan & Partners (TM), I visited New York. In 1949 flying to America was an adventure and I made the most of it. With letters of introduction to engineers in USA and Canada from TM's senior partner, Lawson Scott-White, I was treated most hospitably and returned with three unsolicited offers of employment.

The West beckoned and, happy as I was with TM, I would probably have taken up one of them before long had the partners not responded with an offer that, in the language of today, 'I couldn't refuse'. They undertook to offer me a junior partnership as soon as I became a corporate member of the Institution of Civil Engineers. I remained in London.

During the war, I was educated at Rotherham Grammar School, founded in 1483 as the Thomas Rotherham College by the Archbishop of York and now a Sixth Form College. I already knew that civil engineering was for me. What specially attracted me was the diversity, the individuality of projects not mass produced.

I started at Imperial College in 1944. Wartime college was unusual. The average age was much higher than normal because ex-forces personnel were beginning to come back, some invalided out and some as a result of early demobilization for intending engineering graduates. From 1945 onwards especially, a third or more of a 'set' were of a maturity and commitment altogether greater than the 'green' school-leavers. The mix benefited us all.

Two years later finals were safely navigated followed by a postgraduate year, the first in peacetime. With the pressure of formal examinations removed there was time to study items of diverse interest, within or additional to the syllabus, to read widely and to participate fully in college life. I was living in the Union 'quad', was captain of IC's fencing team and London University's, and doing interesting work academically. Albert Hall was a step away: its classical repertoire became familiar.

Specialization did not occupy much thought but my particular interest was in structures, with the advantage of Professors Pippard and Sparkes in that subject.

Sparkes, who as a young man had trained for a time under George Kirkland, suggested that I apply to TM on leaving college.

In 1947 I started with TM as 'assistant under agreement'. Initial pay was then everywhere a pittance but training was taken very seriously and new graduates were not expected to earn their keep for some time. Like other consulting firms, TM was just getting into its post-war stride. Of the three partners (Reginald Travers Morgan had died in 1940), Kirkland had served in the Army while the older Scott-White and Cecil Pike had been in civilian war work. The practice had been kept going during the war by a handful, including the chief draughtsman, who had been either too old or too young to go to war.

Much learning followed. The rule was 'if you don't know, ask'. I was lucky to work on small jobs for each of the partners and learned much from them, not limited to engineering. The ethos of the firm was greatly influenced by Cecil Pike who attracted respect and affection from everyone. When drafting some specification it was he who passed on the principle attributed to Reginald Travers Morgan: 'Any engineering communication, drawing, specification, or instruction, must not only be capable of being understood. It must also be incapable of being misunderstood.' An ideal aimed at but not always attained. Some years later something had gone wrong on a job in progress. I attended the progress meeting. Various people talked all over the shop, explanations, justifications, running for cover and the like. Pike intervened after some time. 'The first thing we've got to do, gentlemen,' he said, 'is to get the job right!' It was a measure of the man: getting the job right was paramount and, together with the highest quality of service to clients, was what the firm was about.

While learning much at work, I was also teaching theory of structures at evening classes at Battersea Polytechnic (much later subsumed in the University of Surrey). My 'students' (some twice my age) were on their final lap of the external London BSc(Eng), a demanding task. With their commitment and experience the evenings often included dialogue and argument. They demanded understanding and I had to keep ahead of them, learning much more about the subject than I knew when I started. Most pleasingly, none failed this subject during their examinations. Thirty years later I met a leading Malaysian engineer. He reminded me that he had been one of my students; moreover he and others had been most heartened when they heard that their young teacher had become a partner in TM. I was very touched.

I worked to Kirkland on two early projects that were influential. In the post-war period the shortage of steel encouraged the use of the quite new technique of prestressed concrete, and also forms of construction, such as barrel vault shell roofs, which used much less steel than conventional construction.

The footbridge over the Cherwell at Oxford[1] was a very small bridge but the first ever prestressed concrete fixed arch. It received enormous interest, no doubt

---

[1] Design and Construction of a Prestressed Arch Footbridge at Oxford. *Concrete and Constructional Engineering* 45, No 10, Oct 1950.

helped by Taylor Woodrow's Christmas card showing the completed bridge, with all due credits, sent all over the world! Years later I learnt that Yves Guyon, the eminent French engineer, had devoted some attention to it in his masterly textbook on prestressed concrete.

Prestressing had not been applied to statically indeterminate structures in UK and I was on my own as to how to analyse the arch. In those pre-computer days, the calculations were tedious but not intrinsically difficult. In 1998 this bridge was listed as 'of special architectural or historic interest' (Class II) by the Heritage Department (as was another of mine, Winthorpe Bridge over the Trent near Newark, Class II*).

Bournemouth Bus Garage[2] presented different analysis problems. It was originally to have been a fairly standard barrel vault roof under 100 ft span, and would doubtless then have been in reinforced concrete. At a meeting the architect casually reported that the client's requirements had changed: the span would be 150 ft. Shaken, I intervened to point out that in UK such large vault spans had not been done before and prestressing had not been applied at all to barrel vaults. 'Good' said the architect! There followed a lot of study on the design, analysis and construction of barrel vaults.

The analysis was quite difficult. Part of the long calculation required the solution of 8th order differential equations, an 8×8 matrix to be inverted (or two 4×4s). The work had to be formally organized, calculating page by page, each closely-written page being checked at once by a colleague. The noisy Brunswega cylinder hand-cranked calculator worked overtime until we got the marvellous Marchant electric calculator, $10 \times 10 \times 20$ digits and much quieter. It weighed 40 lb and cost (1952) some £400 – not a trivial sum then, equivalent to more than £6 000 now. A pocket calculator today does very much more and is almost given away! Indeed the whole analysis would today be trivial number-crunching on a PC: one up for progress. The building was listed Grade II in August 1999.

A number of such roofs were done, for industrial, educational and hospital buildings. There were also a series of commissions for the War Office, including a large new project at the experimental establishment at Shoeburyness. Years later I would see it from a helicopter as a member of the Roskill Commission, examining a possible site at Foulness for an airport.

You had to be at least 25 to sit the final part of the Institution of Civil Engineers examination. Aiming to become an associate member (now member) as soon as possible, I sat it the day after my birthday in October 1951. Soon after I became a partner of TM (having also passed the Structurals' exam). Fellowship followed in 1959.

In 1952 six months were spent in South Africa, helping our South African partner to establish a small structural practice and also giving a course of lectures on prestressed concrete at Witwatersrand University. These were repeated later at Battersea Polytechnic.

[2] With G.W. Kirkland, Design and Construction of a Large Span Prestressed Concrete Shell Roof. *The Structural Engineer*, 29, No 4, April 1951.

On my return in 1953 there followed a number of small but interesting jobs, accounts of some of which were published.[3] Throughout the 1950s I also became involved in our work in Nigeria. Our first job there, under resident engineer George Mould who was later to become a partner, was Onitsha Market, a trading centre on the Niger in the then Eastern Region. On completion, we were invited to stay for a variety of low-cost road and bridgeworks. It was an interesting time to be working there, both before and after self-government. Our resident partner, Maurice ('Spud') Hayter, built up a substantial and successful team of expatriate and local staff. The young staff we sent out had responsibility much earlier than they would have had at home. Several of them stayed with us on their return and did very well. Robin Wilson, who had joined us specifically to work there, started the first soils laboratory for hundreds of miles around. He later became a partner and my successor as senior partner. We left Nigeria with great regret at the start of their civil war.

In the 1950s in the UK there were few new bridge or road projects being commissioned by the Ministry of Transport and little money for such work. But some cities and county boroughs were keen to start some bridge crossings. We were appointed for Clifton Bridge[4] in Nottingham. In UK it was then the largest prestressed concrete span (275 ft) and the first concrete bridge constructed by cantilevering out.

Over the years other cities' commissions followed, though it was often many years between preliminary report and design/construction: money remained tight and the cycle of economic boom and bust resulted in capital works being put on hold, more than once. I came to know well a number of City engineers: Finch (Nottingham), Minter (York), Wooldridge (Southampton), Campbell-Riddle (Oxford), Burroughs (Cambridge), Anderson (Belfast). Although different personally, they shared characteristics which made them a pleasure to work for. Of high integrity, they were direct and forthright in working relationships, and let you know exactly where you stood. They also knew and understood their local politicians. It was satisfying to work in these cities. Two of the bridges had royal names, Elizabeth Bridge in Cambridge and Queen Elizabeth Bridge in Belfast, opened by Her Majesty, with the unusual distinction of being featured on a Northern Ireland £1 banknote. Bridgeworks commissions later also came from the Ministry of Transport, often for highways being done by county surveyors.

Anna and I were married in 1959, and twins Simon and Paul were born in 1962. Her background, daughter of a regular soldier killed on active service and step-daughter of a diplomat, was beneficial on diverse fronts. She can be both forthright and diplomatic, attributes rarely called upon simultaneously, and socialized skilfully with incomprehensible delegates at international conferences. Civil

[3] Site Strain Measurements. Two papers presented to the *Institution of Civil Engineers' Conference on the Correlation between Calculated and Observed Stresses and Displacements in Structures*, 1955.
[4] (with R.M. Finch) Clifton Bridge, Nottingham. Two papers to the *Institution of Civil Engineers Proc.*, 12, March 1959.

engineers are often away from home and during my 'frequent traveller' years Anna devoted much time to writing, culminating in membership of the Society of Authors.

From the late 1950s onwards TM grew rapidly. This presented a professional problem familiar to all in similar circumstances. Staff, projects, size and complexity of enterprise increased as did my corresponding span of management control. It is satisfying to be looking after more projects, larger projects, and involved with important and topical matters. But this leaves little time for personal contributions to individual projects. Happily there were good colleagues and supporting staff on whom reliance could be placed. Perhaps the most satisfying role for an engineer is that of project engineer for a single project. He can then know everything about the job and it will well reflect his or her abilities, outlook and attitudes.

One of a pair of 'advance' bridges, to be built over the river Weir in Durham before construction of the Sunderland by-pass started, had both drama and lateral thinking. Leon Turzynski, later to become a TM partner, was project engineer for both.

The bridge site was over ancient coal workings at various depths. The drama came when the river rose unexpectedly about 5 ft over a weekend, just high enough to flood some site investigation boreholes! On Saturday night the area chairman of the Coal Board had telephoned the county surveyor at home and told him, I guess on the following lines, '... to stop bloody well flooding my workings...'. There was much anxiety since a build-up of water below could break through to other workings with disastrous consequences. The county surveyor, Basil Cotton, an admirable man whom we held in high regard, had not been involved at all but immediately organized some trucks to dump soil over the boreholes and stop water getting down. We were to meet the Coal Board team on Tuesday.

Long train delays caused us to miss a preliminary meeting with the county surveyor, so we walked straight into the meeting to find some 15 Coal Board people. The atmosphere was tense and the 'London consultants' were due for a caning! Invited to open the meeting, I gulped, apologized for bringing them out unexpectedly, said that the problem had been stopped and with the river level falling would not occur again, that if there had been any foolishness in procedure it had nothing to do with the county surveyor, it was entirely ours and the responsibility was mine. I hoped we could continue the project with the same degree of cooperation and goodwill that it had so far enjoyed. As someone graphically described afterwards, he had '...never before seen the steam go out of a meeting so quickly'. The senior Coal Board representative went through the motions of expressing criticism and warnings but everyone could see that this was routine. The meeting was quite short, we heaved a sigh of relief and stayed for tea.

At the bridge site the curving river was athwart a geological fault. One side was over old coal workings but the other side had never been subjected to mining. The bridge had to sustain foundation movements and rotations caused by the mine workings. That such workings were only on one side made it more rather than less difficult.

119

We were well into the design and still not entirely happy with what was emerging. Measures dealing with subsidence would add substantial cost. At periodic team meetings we worried at the problem to no avail. Then at one meeting someone said, in exasperation, '… rotten place for a bridge. Can't move the road without new statutory procedures, pity we can't move the river.' Suddenly, we started to think. Why not move the river? Estimates for straightening the bend, thus putting the bridge beyond the fault on solid ground, quickly followed. The savings from a straightforward bridge, built before river diversion on conventional foundations in a field, would make it the more economical solution! Since the Ministry would pay the cost of the diversion the River Authority was happy to agree. Later and off the record, my opposite number told me 'we've been trying to straighten out that bend for years but could never find the funds for it.' After that it was plain sailing!

Our first motorway appointment, in 1962, was for the M23, then known as the Brighton motorway though terminating at Pease Pottage and directed to '…a suitable terminal in South London…'. With the abandonment of the proposed London motorway box later it terminated south of Purley. Initially we had to report recommending a route and then take the whole of it through the many administrative processes to reach a public inquiry. We broke new ground in approach and our report reflected this. In particular, we were very conscious that our work would be severely tested at the 1967 public inquiry: there were plenty of complex problems and no shortage of well-equipped objectors.

Our approach was later referred to as 'planning by alternatives'.[5] Different possible routes had particular advantages and disadvantages. Some were capable of being costed in money terms, some not. Cost was only one consideration. We made a basic decision: all 25 feasible alternative routes would be developed, described and evaluated to the same degree of detail. Thereafter a formal process of pairwise comparison should lead to the preferred route in an explicit and transparent way. Commonplace now, quite new then.

Whilst researching history of transport in the area we found that the world's very first motorway proposal was the private bill for the 'London and Brighton Motor Way', lodged in 1906 but subsequently withdrawn by the promoters in view of much opposition. It was a remarkable proposal, extraordinarily far-sighted.

Following a three-week public inquiry, the inspector recommended the published line. We organised the arrangements at the inquiry, some of which were new, and they worked well. We were gratified when the Ministry asked us to write them all up by way of a public inquiry manual. This became an in-house document and went to several editions.

Other motorway appointments followed. Working on them involved us in traffic evaluations, simple initially but becoming increasingly demanding. Later they be-

[5] Motorway Route Location Studies. Main paper presented to Town and Country Planning Summer School, Keele, 1966.

came transport studies (all modes of transport). We carried out such work in historic towns, York, Cambridge, and also for areas, e.g. the South East Regional Plan. Transport planning inevitably led to involvement in land use planning. This area became very much the province of TM partner Chris Holland. He had joined us as graduate engineer and had become a member of the Royal Town Planning Institute. We worked together on some major studies.

The 1960s onwards was also a time of committees, inquiries and working parties both for the profession and within the aegis of the Civil Service. The first of the latter was an interdepartmental committee on the work and role of civil engineers in the Ministry of Transport. I was the only external member and learnt much about the Service. Our report was favourably received and shaped the development of the engineering function in the Ministry throughout the 1960s.

In the Urban Motorways Committee (1969–72) I like to think that I played a part in getting an increase in the compensation paid for compulsory purchase of homes for highway purposes. In a large committee there were just two or three of us who felt keenly that householders' surplus ought to be recognized by an extra payment. Normal market value (the price for willing buyers and sellers) plus moving expenses does not adequately compensate someone whose home has to be compulsorily purchased, yet it was then the rule. Several of the committee members argued strongly against our notion but in the end we prevailed. The shortly following Act allowed for a 'home loss' payment.

In 1968 came the invitation to be a member of Lord Roskill's Commission of Inquiry on the Third London Airport. Its task was to make recommendations on the timing of the need for a four-runway third London Airport and its site, taking everything into account, and to carry out what was then the largest and groundbreaking cost-benefit study. It proved an absorbing and taxing experience on a highly controversial issue.

For about 30 months we met every Friday and there were additional meetings and local public inquiries during this time. For four months we sat four days a week during the final public inquiry, being addressed by about 40 leading members of the planning bar. It was public service, there was no remuneration. We got to know each other well and I learned a lot from our superb chairman. The state-of-the-art methodology must have been strange to him at first, but his keen mind and very agreeable personality enabled him to lead us all, commissioners and research team, in a masterly fashion. We became very fond of him. Our friendship continued until his death.

We rejected Foulness as a site. Events have subsequently proved that right. But no one foresaw, or could foresee, the effects of two later developments. One was the introduction of much larger wide-bodied aircraft, the other advances in air traffic control and its instrumentation. The combined result was that in UK restricted space conditions, runway capacity is not yet the main constraint on airport throughput. Passenger handling and terminals have become the key. All airports have advantages but also serious environmental disadvantages. As far as can be seen ahead there will be no four-runway airport in the UK.

The team's technical work was put under a microscope at the main inquiry without being seriously impaired. The fact that Government did not accept the admittedly controversial recommendation was disappointing but did not in itself cause the Commission much heartache. Judges are well used to being overturned on appeal! But what did trouble us was the opinion that we were simply led by the results of the cost-benefit study. That was just not true and we had said so. No such study can persuasively value all the factors in monetary terms; it was a guide to our collective judgement on the brief we were given.

For myself, I found the experience professionally rewarding, highly educative and, in the main, agreeable. Alan Walters, the distinguished economist who was a member of the Commission, mentioned in acknowledgments of a later book that Roskill and myself were '…naturally born economists…' but went on immediately to say we would consider that a congenital defect! Certain it is that I knew more about economics, law and lawyers when it ended. A short account of airport siting is in [6].

There were many brickbats after the event, leavened by a few bouquets. 'Goldstein was to emerge as one of the most formidable members … with a firm grasp of the whole of the complicated field it was surveying and an ability in questioning witnesses which any lawyer might have envied' (*A Sadly Mismanaged Affair – a Political History of the Third London Airport* by David McKie, Croom Helm, London, 1973).

From the sublime to the …? Shortly afterwards I found myself in a unique position where cost-benefit was both relevant and instrumental. It concerned a proposed interchange for the M25 at Westerham, Kent.

We had moved house from London to near Westerham in 1966. About three years later the South-East Road Construction Unit published statutory plans for the proposed interchange. A local resistance group was formed and asked me to join. I declined. Our house was over two miles south of the road, over the brow of the secondary ridge of the North Downs, well out of sight and sound of the road. Our house value would increase with an interchange rather than the reverse. I was then asked to accept a professional appointment from the resistance group to advise. I could not do that. The Unit was our client on other work and we were doing day-to-day business with them. The response to my explanation was frosty. My professional attitude was rather counter-productive if we wished to become a member of the local community (we still live there). So I recommended a good engineer to them and undertook to examine only the traffic and economic evaluations of the proposal. I expected to find them unexceptional and having done that could gracefully withdraw without engendering hard feelings.

However, when examining these aspects I found all far from well. So I called on the senior man at the Department of Transport, explained the position and sought his advice. Without a moment's hesitation he said that if I thought the proposal was wrong I should become a formal objector as an individual. He was

[6] Criteria for Siting of Major Airports. Paper 21. *4th World Airports Conference*, London, April 1973.

kind enough to say it might prevent them doing something unwarranted. So I became one of the objectors and carried out technical work ready for the inquiry.

I appeared at the inquiry as principal, advocate and professional adviser rolled into one. We never found out the inspector's opinion! Late through the inquiry the Department required copies of all the proofs of evidence for the then engaged counsel. Spare copies were no longer in stock. They inadvertently ran off copies of the inspector's and passed them to counsel. During the inquiry it emerged that these copies contained some inspector's notes. The roof fell in! A little later the Department withdrew the draft order and that was the end of the proposal.

The view now is that the M25 has too many interchanges permitting more local traffic to add to the overload. So the outcome was right, whatever the reasons.

TM started using mainframe computers in the 1960s. By the mid-60s I had some 'hands on' experience of some universities' Feranti Pegasus machines, and found it fascinating. (Little did I know then how large a role computers would play in my retirement.) In 1969 we got our own IBM 1130, lifted in by crane through a window. By today's standards its capacities were laughably trivial but a trio of TM engineers made the machine sing and dance in a way that astonished IBM engineers. They were Brian Gee, ex IBM, who was in charge of the machine and systems, including probably the first management accounting programs used by consulting engineers; Keith Johnson, a transport analyst who was especially skilled in array and matrix operations; and later Dr Graham Anderson, a structural engineer especially capable in finite-element analysis. By use of overlaying and virtual memory techniques the capability of this early computer and its upgrades was enormously enhanced.

With continued growth of the firm I was fortunate in two of the staff who worked closely with me and whose help and understanding must be acknowledged. One was Lt. Col. Dick Painter, our chief accountant. Fast growth has to be financed and puts pressure on accounting operations, increasingly as methods are changed by early computer technology. Painter had left the army early to qualify as a chartered accountant. He did a marvellous job for us and managed a dedicated accounting staff extremely well.

The other was Miss Pat Sandilands, my superb secretary. She deserves much more than can be said here. A senior partner's secretary has to blend a judicious mix of protectiveness and accessibility for her chief. She is also a legitimate source of initial advice for staff as to how he might react to this or that. She becomes a very important person indeed in the organization and has to have the temperament, attitude and style to respond to that.

Probably because of my experience concerning the third London airport, we were engaged to carry out a location study for the second Sydney airport, including a comprehensive cost-benefit study, and make recommendations. I had not visited Australia before, found the country fascinating and the project absorbing.

A joint Federal-State Committee had been charged with the task and we occupied the same role vis-a-vis the committee that the research team had occupied with respect to the Roskill Commission. In fact some of the research team came to work for us on the project. The committee's chairman was Jim Harper, under-secretary at the Department of Aviation. A second airport site would of course be highly controversial.

It was to be a major study lasting two to three years with an interim stage after about a year. I visited Australia every three months or so, *inter alia* to attend and report at formal review meetings of the committee. Working in Australia was agreeable and the atmosphere wholly encouraging towards trying new ideas. Harper was a most engaging personality, superb as a client, and became a close friend. We adopted the Roskill team methodology and developed it further in several areas. Our first recommendation was that rather than a second site, a second parallel runway for the existing airport would be by far the most economic solution and would preclude the need for a second site for many years, though such a site should be reserved. At the interim stage we put forward a short-list of sites for detailed study.

There was a change of government and Whitlam's Labour Government decided not to proceed with the detailed study. Instead, the Prime Minister announced a site near Galston. A strongly conservative and prosperous area, it had been a doubtful runner and had not made it to our short-list. We were all convinced that a Galston airport could never happen, nor did it. Some months later the Prime Minister announced that it would not be necessary to have a second airport there.

The fate of our first recommendation was interesting. The noise envelope of a second parallel runway would affect two important constituencies and Government found it impossible to proceed with it at that time, for environmental and political reasons. Traffic management optimization maintained acceptable conditions for a while, but eventually, as congestion increased, something had to be done. The second runway was built about 20 years after our recommendation!

The premature end to our study was disappointing. We were treated sensibly by our client and for a time were engaged for other aviation-related work. As that ran out, we were encouraged to set up a permanent firm in Sydney to advise on all modes of transport. This we did and first under Chris Holland and then Piers Brogan the company prospered, carrying out much state-of-the-art professional work. As partner accountable for that new enterprise I continued as a regular visitor to Australia, covering also Malaysia where we were working, until my retirement.

It is curious how projects starting in difficult circumstances often go well, not least because everyone concerned, knowing the difficulties, does his best to help others. One such example was the Itchen Bridge at Southampton,[7] one of the best examples of local authority governance I have encountered. There was a

---

[7] (with W. Fox) Itchen Bridge, Southampton – Design and Construction. *The Structural Engineer*, 55, September 1977.

long-standing promise by the City that the ferry to Woolston should be replaced by a permanent crossing. The high-level bridge went to tender at the worst possible time, with the 3-day week, wage restraint, strikes, scarcities of supply, rising costs etc. The received tenders were naturally heavily qualified, in a way that would be unacceptable in normal circumstances and would lead to delay and re-tender. But the City's responsibility was about to be extinguished by local government reorganization. The county taking over would not have the same priorities. We were instructed to negotiate with the successful tenderer as best we could, so as to minimise the qualifications.

I attended the last committee meeting with misgivings. They wanted to build the bridge and had funds ready for it. If asked my professional opinion I would have to counsel delay in view of all the difficulties and the potentially rather open-ended contract commitment. Happily, the chairman never asked me that question. Asked whether it would be possible to construct the bridge with the contract to hand and supervise it accordingly, the simple answer was 'yes'. The committee and the council approved the contract, it was signed just before reorganization, and work started. In the event the job went well under project engineer George Teller, and the downside scenarios did not occur. The bridge was opened by Princess Alexandra, who had also opened our Clifton Bridge in Nottingham 20 years before.

Advisory work, both on planning and transport matters and on construction defects and contract disputes ('forensic engineering') increasingly occupied me. There was arbitration work and an interesting conciliation.[8] I was even visited by Scotland Yard on one occasion! They were interviewing all engineers of jobs done by a contractor during the previous few years. On a current job, of which we knew little, one of his employees was alleged to have acted improperly in connection with some claims.

The two officers from the Fraud Squad who interviewed me knew their stuff. They were interested in engineers' procedure for processing claims. They were puzzled by the commonplace small proportion of claimed sums certified. They pressed me on whether contractors would submit claims even though they did not think they were entitled under the contract to all they submitted. I could not answer that authoritatively but repeated to them what a doyen of the contracting industry had told me many years before. 'It is a problem' he had admitted, 'the contract is very complex, we don't know which claims will find favour, sometimes an engineer rejects what we are sure about and accepts something where we're not certain, so you can understand a contractor throwing everything in'. The two officers were not impressed. 'If a contractor submits a claim containing anything at all for which he does not genuinely believe he has an entitlement in law, we have a name for that' one of them said. On my enquiry he said something, I wish I had written it down, on the lines of misrepresentation for pecuniary advantage. 'And

---

[8] The Case for Conciliation. *Building*, 25 May 1979.

our name for that' he continued 'is Fraud'! The fact that not a penny would be paid which the engineer did not certify as an entitlement for the contractor made no difference to their views. Having taken a statement from me they departed amicably.

Membership of various advisory bodies continued. For most of the 1970s I was first a member and then chairman of the Planning and Transport Research Advisory Council for the Departments of the Environment and Transport. Arising from that work, and also our work in the office, I became increasingly concerned about certain methods of evaluation and comparison of projects.

There are some innate fundamental difficulties about such methods for public works. Many of the factors to be compared have no common measuring scale, certainly not a scale of money. They are referred to as 'intangibles'. We have to identify, measure, and value them. We have done quite well with identification and can usually measure an effect in some way. But valuation remains elusive.

Another problem is that the 'gainers' are spread very thin, i.e. there are vast numbers of people who each get modest benefits, whereas the 'losers' are much fewer and losses are individually much higher. So losers naturally mobilize against the project; gainers are often the silent majority.

Perhaps the most serious problem relates to the aggregation of individual opinions or preferences. When there are three or more mutually exclusive alternatives containing many intangibles, all that individuals can do is to rank them in order of preference. These 'ordinal' values cannot be reliably aggregated to produce a 'community opinion'. Accounts of the problem are in [9] and [10]; the last part of [11] summarizes the position.

What troubled me was that many planners and engineers tried to overcome these problems by operating a sort of 'numbers game'. We've all done this at some time. Points are awarded for this and that and are then simply summed, or subjected to other arithmetical processes, as if the result had formal and persuasive meaning!

Preference lists provide no measure of the 'distance' between the items. But that is not the only reason for the difficulty. If asked to put it in a nutshell, I would rely on some aphorisms. *De gustibus non disputandum* is learned, 'there's no accounting for tastes' is equivalent, 'one man's meat is another's poison' is more commonplace, and the Yorkshire 'there's nowt so queer as folks' is more homely. But they all touch on the analyst's formal truth, 'the interpersonal comparison of utilities is invalid'.

---

[9] Highways and Community Response. Ninth Rees Jeffreys Triennial Lecture, Royal Town Planning Institute, December 1975.

[10] The Expert and the Public: Local Values and National Choice. Seminar by Department of Planning and Department of Philosophy, University of Florida, Gainesville, 19–23 March 1986. Published in *Business and Professional Ethics Journal*, 6, No 2, Summer 1987, University of Florida.

[11] Travel in London: Is Chaos Inevitable? Sponsored lecture. London Regional Transport, 16 November 1989.

Though these numbers games still go on, I was able to stop one or two of them.

A variety of work continued to come in. Work in building structures such as hospitals, offices, industrial buildings, were regular commissions being carried out under the leadership of TM partner Keith White. Another partner, Turzynski, found himself increasingly in demand in forensic engineering.

By 1985 I had been senior partner for 13 years and my thoughts turned to departure. There were more years to go before our partnership agreement required retirement and nobody was 'pushing' me. But I'd been there a long time and when the same policy and management questions come round for the third time it tells you something (other than that the questions seldom change). Also, there were things I wanted to do for which hitherto I had been unable to devote the time. So I retired and became an individual part-time consultant for about seven years advising both the TM Group, for whom I became non-executive chairman for two years, and other clients. It was a very sound and felicitous decision and my advice to others reflects my experience. 'Go early rather than late; go whilst you are keen to do other things and have the energy to do them; go when there are good prospects for your successors for at least the following few years.'

A short chapter like this cannot properly describe over 40 years' professional life. So, looking back, how should I conclude?

I enjoyed being a civil engineer. It is a very worthwhile career, the training for which fits anyone for a broad range of opportunities. One of the attractions of civil engineering for me when young had been the diversity of possible activities. I certainly had that: structures, bridges, roads, traffic, transport, planning, forensic engineering, with concomitant skill development in economics, management, operational research, some law and a little political science. Work in various parts of the world, some public service and a fair number of demanding challenges, let alone controversies, prevented more than due share of routine.

But many opportunities arose by chance, by coincidence, or by random combination of circumstances. So I was lucky or, perhaps in the right place at the right time, had my share of luck. Not long after retirement a senior official asked how I found my new role. I observed that amongst the things most welcomed were the absence of the tyranny of the diary, the independence when working at home provided by the computer and the fact that professionally I now did only what appealed to me. 'But haven't you always done that?' he smiled. I realized that I had indeed been fortunate.

Napoleon favoured 'lucky generals' but luck alone is not enough; there is also commitment. The mother of an outstandingly successful businessman was interviewed on his 50th birthday. She was asked 'Would you not agree that your son has been very lucky?' 'Yes,' she acknowledged, 'but it's a funny thing, the harder he worked the luckier he got!'

Alfred Goldstein, CBE

BSc (Eng), DIC, FCGI, FREng, FICE, FIStructE, FCIT, FIHT

Born on 9 October, 1926. Education: Rotherham Grammar School; Imperial College, University of London. Consulting Engineer; Partner, R Travers Morgan & Partners, 1951; Senior Partner,1972–85; Chairman TM Group, 1985–87. Responsible for design and construction supervision of major road and bridge projects and for planning and transport studies. Projects include Clifton Bridge, Nottingham; Elizabeth Bridge, Cambridge; Itchen Bridge, Southampton; M23 Motorway; Transport consultant to Govt SE Jt Planning Team for SE Regional Plan; Belfast Transportation Plan; London Dockland Redevelopment Study; Site location and cost/benefit study for 2nd Sydney airport for Australian Government.

Appointments include membership of Building Research Board, 1963–66; Civil Engineering EDC on Contracting in Civil Engineering since Banwell, 1965–67; Baroness Sharp's Advisory Committee on Urban Transport Manpower, 1967–69; Commission of Inquiry on Third London Airport, 1968–70; Urban Motorways Committee, 1969–72; TRRL Advisory Committee on Transport, 1974–80; Committee on Review of Railway Finances, 1982; TRRL Visitor on Transport Research and Safety, 1983–87; Chairman, DOE and DTp Planning and Transport Research Advisory Council, 1973–79; Board Member, College of Estate Management, Reading University, 1979–92.

Publications: Over 60 papers and lectures including: *Highways and Community Response*, 9th Rees Jeffreys Triennial Lecture, RTPI, 1975; *Environment and the Economic Use of Energy*, Plenary Paper, Hong Kong Transport Conference, 1982; *Investment in Transport*, Keynote Address, CIT Conference, 1983; *Buses: Social Enterprise and Business*, main paper, 9th annual conference, Bus and Coach Council, 1983; *Public Road Transport: a Time for Change*, keynote address, 6th Australian passenger transport conference, 1985; *Private Enterprise and Highways*, National Research Council Conference, Baltimore, 1986; *The Expert and the Public: Local Values and National Choice*, Florida University, 1987; *Travel in London: Is Chaos inevitable?*, LRT, 1989.

Management Experiences

Sir Neville Simms

My career in the construction industry began in 1966 and from that moment to this, whether on site or in the office, five important objectives have been an integral part of my working life. I have always aimed for excellence through a commitment to getting things right first time; I have sought to bring the best team together for whatever I have been doing and, I hope, I have provided that team with strong leadership; I have endeavoured to manage change sympathetically and I have used the power of negotiation to achieve the desired results.

All of the contracts I have been involved with, from the very earliest days in Scotland through to the Channel Tunnel, have provided me with many opportunities to develop my personal and professional skills and, in the main, to achieve these objectives.

After receiving a first class honours degree in civil engineering at the University of Newcastle-upon-Tyne, I joined consulting engineer Ove Arup & Partners and quickly became involved in designing and supervising the construction of schools, hospitals and a wide range of building projects. My very first experiences of work were in a design office and I quickly came to understand that working there, in the real world, is significantly different from working at university, attempting to pass examinations. In the commercial world no marks are awarded for approaching the subject correctly or for having the right ideas. Marks are only given for getting the answer right. After all, you can imagine a real-life situation where you are designing roof trusses – one of the very first jobs I was given was a space frame roof truss – and you calculate the stresses wrong by a factor of ten. A simple mistake at university, which, in my day, would probably have been looked upon reasonably kindly by the lecturer; right idea, right approach, good marks, silly mistake at the end. In real life, of course, getting the design right really matters and there could well be a collapse if something as simple, yet as fundamental, as calculating the stresses was done wrongly.

When I moved out onto site in 1969, the lessons in both life and in management continued. Perhaps one of the first was the realization that theory is not always put into practice; nonetheless the result can still come out all right. I remember walking the very first section of earthworks that I was responsible for with the chief engineer of the contractor for whom I was working and comment-ing to him that the roller being used was neither the best, nor most recommended, type of roller for boulder clay in the central belt of Scotland, where I was at the time. A sheep's foot roller would have had higher compactive energy than the smooth wheel roller that was being used. He very quickly pointed out to me that the best theoretical piece of equipment was not always to hand and you had to use the equipment that was available: perhaps more a lesson of life than a lesson in engineering.

Later I moved from the earthworks of that first job on to bridgeworks and I was sent out into the middle of a field where there was a small river, not much more than a ditch really, running through on the line, a long diagonal line, of what was going to be a motorway bridge. By the time I got there the craftsmen, joiners and steel fixers were already in place and had taken all the coat hooks in the hut which I had to share with them. There were no concessions to management and it was quite clear that I was the junior member of the team even though it was me who was going to set out the bridge and make sure that everything was in the right place at the right time. They were the skilled craftsmen, I was very much the junior engineer. By approaching the work in a confident yet diplomatic way, I was eventually able to win their confidence as they came to realize that the accuracy of my work and my ability to produce materials, for them, at the right time could enhance their bonuses. So from starting off very much as the laddie who made the tea, I ended up rather more as the team leader by the time we had finished that small bridge; which every time I pass over it fills me with considerable pleas-ure and pride as I remember those early days; and I still see and chat to some of the men I worked with, on site, when I am occasionally able to leave the office and visit a project.

My experiences on site over the next 10 or so years demanded both innova-tion and resourcefulness. The innovative use of second-hand materials, for exam-ple, timber for support shuttering, which was considerably cheaper than new tim-ber, and the use of old brick demolition rubble, rather than new crushed stone, saved cost and was, in hindsight, a more environmentally sustainable way of work-ing. The decision to produce my own materials locally by opening up quarries and borrow pits, rather than having to truck large amounts of material by road, was more economic and also avoided disruption in the local community.

At this time, the 1970s, we were deep into union problems and, although these are not problems that many people regularly face today, negotiating with unions is something that both strengthens and broadens one's character. Being trained as an engineer I was used to dealing with facts and figures and basically being logical. I think it is fair to say that the unions tended to have a worldly-wise approach to the issues under discussion and this coloured every aspect of the negotiations, thereby making it extremely difficult, and personally frustrating, to deal with them in a practical sort of way.

The extension to Drax Power Station, where in 1979 I took responsibility for Tarmac's involvement in the main foundations contract, was one of those projects where the unions had historically almost held a whip hand over the contractors. In this environment it was very difficult to get a fair day's work for a fair day's pay.

Following years of disastrous industrial relations a management committee, including the client, contractors and suppliers, was set up for the whole site. I played an active part and, despite a continuing difficult industrial relations climate nationally, it is a credit to those involved that the project was delivered earlier than programmed and at less than the original budget. The tide was turning back in favour of positive management and leadership.

Another major contract, where I was project manager, the Stoke D Road contract, was the largest road contract (in money terms) ever let in the UK, and the largest urban project attempted up to that time. It got tangled up in local union politics, particularly concerning the election of union officials. This had nothing to do with the contract but certainly the issues impacted on the contract and had to be handled with diplomacy, calling for patience, skill and, in the negotiations, an understanding of just what the real objectives of those involved were

Among the highlights of my final years on site was my involvement in the Thames Barrier, one of the largest non-oil projects in the UK at the time, and a job which had its share of union difficulties. But perhaps, in this instance, these were outweighed by the technical difficulties of creating huge holes in the River Thames thirty metres below the surface of the water, for the cofferdams for the massive piers. I was appointed to the joint venture contractors' management board about two-thirds of the way through, at a time when union relationships were still very tense and there was a real danger that the project would not be finished and that the contractors, who had accepted a new form of target contract, would lose a lot of money. Fortunately we didn't and I am pleased to say that the project was delivered ahead of its newly negotiated delivery time and inside its targeted cost, which meant the project was, in the end, profitable for those involved.

One of the interesting aspects of the construction industry is that young men and women get to run their own quite considerable empires very early on. A site, especially a big, remote civil engineering project, can be very much like a small township with all that that demands, in providing leadership, direction and motivation to huge management teams and labour forces.

Choosing the right people, therefore, for the secondary and support roles both in the office and in the field, as well as the key management roles, is a vital ingredient for a successful project. It is my opinion that by the time a young site manager leaves the site, he or she has gained considerable skills associated with managing the commercial and human aspects of the job – whether a profit or a loss is being made, how to control costs and how to direct a labour force that might, in many cases, be made up of casual labour – many of the general management skills needed to run an independent business. They have also probably learned to be mature and philosophical about the things that cannot be influenced, like the weather, whilst all the time endeavouring to minimize the impact.

By the early 1980s I was no longer site-based and, apart from a 'return visit' as the UK co-chairman of the Channel Tunnel Joint Venture (but that's another story), I moved into general management running ever larger businesses within Tarmac Construction. I was also developing a growing interest in how the construction industry worked. As chairman of the National Contractors Group in 1989, I led and promoted a study of four particular areas of the industry: structure, education, research and development and image, called *Building Towards 2001*. The then Prime Minister, Margaret Thatcher, kindly wrote the foreword and over 30,000 copies of the publication were requested and distributed. The ideas on education and research and development were quickly taken up, despite the growing recession in the industry, but structure took longer to develop. However, with the help of the Department of the Environment, and in particular, Sir George Young, the then Minister of State, Sir Michael Latham looked at structure and the unnecessarily adversarial nature of our industry, resulting in his report, *Constructing the Team*, which was published in 1994. Since then, slowly but surely, many of the recommendations have been picked up as the industry has changed to meet the challenges of the 1990s.

To complement those efforts I worked, for nearly ten years, with a number of senior industry figures, to create 'one strong voice' for our industry, an industry which produces 8% of the UK's gross domestic product and employs over 1.4 million people. Our efforts finally came to a successful conclusion in 1997 with the creation of the Construction Confederation which is now able to speak with a much stronger mandate on behalf of the construction industry.

But back to Tarmac where, in 1992, I became group chief executive some 22 years after I joined the company. By then roughly half of my time had been spent in direct control of sites and half had been spent in the office, as a business manager. It is, however, unlikely that anything could have completely prepared me for the role of taking over as the chief executive of Tarmac plc in the early months of 1992.

The construction industry was deep in recession at that time and the Tarmac Group, which had expanded very rapidly in the 1980s, was faced with collapsing markets, deteriorating performance and a rapidly increasing level of debt.

There was need for consolidation, re-orientation, streamlining and re-balancing of the portfolio of businesses to meet the challenge of the markets that were expected in the first half of the 1990s. The key focus of my early attention was on a rapid reduction in the level of debt. As a result of a mixture of offensives, firstly through loss elimination, then through down-sizing, next through business disposals and finally by the addition to shareholders funds through a rights issue, the Group's gearing was eventually brought under control.

But to make real changes in a very large group you have, of course, to change the key people; people who can often be wedded to outmoded systems and a style and a culture that are very often a hindrance at a time of rapid change. I started to implement change at the very top of the Group, with the board of Tarmac, which had numbered fourteen when I took over. It was quickly shrunk down to seven, three executives and four non-executives, since the actions that were needed required some very closely managed decisions.

However, as the debt crisis started to come under control, so we were able to involve more and more people in choosing the future direction of the Group, which at that point in time, was able to move from the 'sick bed' to the 'rehabilitation' unit. We were able to be more proactive and less reactive.

Our initial strategy was to focus on three areas of operation, construction services, private sector housing and quarry and heavy building products rather than the seven divisions that had developed in the 1980s. It soon became clear, however, that the investment demands of even three activities could not be properly satisfied.

It was a very difficult time as the senior management team agonized over which activities suited the Group better, bearing in mind that we were the largest practitioner in the UK in all three areas. Finally a decision was made which led to the disposal of the housing business and the strengthening of the mineral and the contracting activities. We then negotiated and completed the £600 million business swap with Wimpey between September 1995 and March 1996. At a stroke, Tarmac was able to transfer quickly and tidily £300 million of capital employed in the volatile and cash-hungry private sector housing market into the more stable heavy building materials and construction services business activities, which, by the way, was a reflection of the Group's history for 70 of the 90 or so years that had passed since its formation.

With the decision made to run the two business streams, the culture of the organization, which had been changing quite significantly over more recent years, was tackled even more vigorously. The Group was freed to pursue new values, values associated with customer focus, quality and innovation. It was also able to step up its people-orientated programme to help it to achieve its aim of becoming 'world class'.

And again, some of the lessons that had been learned in my very earliest days on site came to the fore. It is all very well to have a 'top down' approach to cultural change but its not until you see the messages coming back towards you from deep inside the body of the organization that you know that you are starting to succeed. I am, therefore, personally, closely involved in our continuous performance improvement programmes. We call the initiative Target 2000 and it encompasses a wide range of activities aimed at drawing the best out of every employee in the Group, helping Tarmac to achieve its stated objectives of providing leadership in the markets in which it competes; having a world-class reputation; employing a workforce its competitors envy and its clients demand; and maximizing shareholder value.

And so to conclude, I believe that the construction industry offers enormous scope for self-initiating people with potential who:
- understand the need to lead;
- understand the need to involve all of their team, even though they are designated the leader;
- can develop good negotiating skills, and
- have good personal qualities, particularly those associated with integrity and fair play.

133

Finally, I believe that there is an exciting and satisfying career for the young engineers of today in the construction industry provided, they have an ability and willingness to involve themselves with others and, most importantly, the drive to aim for excellence in all that they do.

Good luck to you all.

Sir Neville Simms

BSc, MEng, FREng, FICE, FCIB, CBIM, MIHT

Born in Glasgow, Sir Neville now lives in the West Midlands. He was awarded a knighthood for services to the construction industry in the 1998 New Year Honours List. He was appointed Doctor of Technology by the University of Wolverhampton in 1997.

He joined Ove Arup & Partners in 1966 and in 1969 he moved into the contracting side of the industry with A.M. Carmichael Ltd of Edinburgh and, in 1970, Tarmac Civil Engineering. He gained wide experience of contracting before being appointed Director of Operations for Tarmac National Construction in 1980, and a Director of Tarmac Construction Ltd the following year. In 1984 Sir Neville became Joint Managing Director of Tarmac Construction Internationaland the following year Managing Director of Tarmac Construction's regional operations, responsible for projects carried out throughout the UK. In January 1988 he was appointed Chief Executive of Tarmac Construction Limited and in March 1988 he was appointed a Main Board Director of Tarmac PLC. On 14th February, 1992, he was appointed Group Chief Executive of Tarmac plc, the UK's largest heavy building materials and construction services group and in March 1994 took on the additional role of Deputy Chairman.

He is Co-Chairman of the Members Assembly of Transmanche Link, the contractor which built the Channel Tunnel and President du Conseil de Surveillance de l'Entreprise Nicoletti in Nice, France. Sir Neville has served as Chairman of the National Contractors Group, a member of the Overseas Project Board (DTI) and the Environment Task Force of the Building Employers Confederation, council member of the CBI and Chairman of the Major Contractors Group. Between 1993 and 1997 he served on the Private Finance Panel,and has been Chairman of Balsall Heath Birmingham Employers Forum, under the auspices of Business in the Community, since 1994. He is also a non-executive director of Courtaulds PLC, and the Bank of England and a governor of Stafford Grammar School.

135

Construction sites are the best teachers of construction engineering

Koichi Ono

*L*et me introduce my career briefly. I completed a bachelors then a masters degree in civil engineering at Kyoto University in 1967, then went to Canada and studied at the University of Calgary from 1967 to 1969, spending most of the time improving my English and catching up with course work.

My supervisor, Professor R.H. Mills, accepted a job offer from the University of Toronto in 1969 and asked me to start a PhD course at the University of Toronto. I completed my PhD in 1972 in the field of fracture mechanics using the finite element method. Although I had the chance of a position from the University of Toronto, I accepted an offer from a Japanese construction company since I wanted to study practical engineering rather than continue academic work. I was fortunate in learning so much during my five years in Canada, and I deeply appreciate my professors and classmates. I still have a good relationship with many of them.

I spent 24 years with the construction company. In 1996, I was promoted to Professor in the School of Graduate Study at Kyoto University. I was very fortunate to have the opportunity to work on many big projects and engineering problems while with the construction company.

I would like to introduce some engineering lessons obtained through my experience.

Open excavation

In the mid 1970s, I became chief engineer for the construction of a sewage plant. The key part of this work was the excavation of an area 70 m × 100 m × 22 m deep, in very loose and soft ground with a very high water table. A reinforced concrete wall was used as the earth retaining structure. It was designed by a consultant and the required thickness of the wall was determined to be 0.9 m.

In Japan, most construction companies check the design of a structure after award of contract. The required thickness of the wall turned out to be 1.0 m by my calculation. The difference in thickness was caused mainly by different assump-

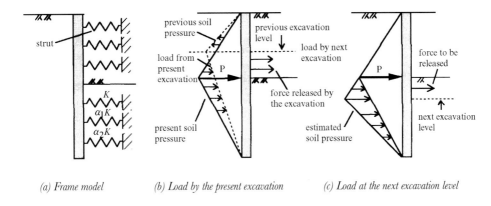

(a) Frame model   (b) Load by the present excavation   (c) Load at the next excavation level

FIGURE 1 *Frame analysis of a retaining wall.*

tions for the soil pressure. It is, in general, not easy to seek a change in design from clients. I proposed, therefore, to monitor the wall during excavation instead of discussing uncertain values for the soil pressure. I developed the following safety check method using a monitoring programme.

Figure 1(a) shows a structural model for an earth-retaining wall, where the wall is modelled as a plane frame structure with both strut and supporting ground represented by springs. Calculation of displacement and bending moment of the wall was done using ordinary frame analysis based on the following assumptions.

The spring constant for each layer of the supporting ground is assumed to be proportional to the N-value in a sandy layer and to the cohesive strength in a clay layer. Therefore, the spring constant for each layer can be represented by the present first layer of spring constant K. In this analysis, K is treated as unknown.

The load acting on the wall, which is the resultant of the active and passive soil pressures, is assumed to be of triangular distribution with a maximum load intensity P at the current excavation level as shown in Figure 1(b). Calculation of the displacement and bending moment of the wall is done incrementally depending on the excavation levels. The force released by the excavation is assumed to be the axial force applied in the spring of the newly excavated soil. Only the load intensity P at the present excavation level is unknown with respect to the load. Reduction of the rigidity of the wall due to the bending cracking is also taken into consideration.

The spring constant K and load intensity P must be known in order to accurately determine the displacement and bending moment of the wall. They are determined so that the calculated displacement of the wall matches the measured one using, for example, the least square method.

Prediction of displacement and bending moment of the wall after the next excavation can be done using the newly determined P and K. In this prediction, the intensity of the soil pressure at the next excavation level is assumed to be the extrapolation of the present load, as shown in Figure 1(c). Then, the displacement and bending moment after the next excavation can be evaluated before excavation takes place.

The predicted displacement and bending moment of the wall and axial force in the struts are used to evaluate the safety of the earth retaining wall. Appropriate action can then be taken prior to the next excavation if necessary.

Several reinforcing methods of the wall were prepared for the case where the wall displacement was estimated to exceed the design value. In fact, when the excavation proceeded to about two-thirds of the required depth, the prediction indicated that the wall displacement would be exceeded significantly at the final stage of the excavation.

It is, in general, not very easy to reinforce an earth-retaining wall during excavation. At this site, I adopted two reinforcing methods: ground improvement of the excavation side and pre-setting of the struts at the next excavation level by digging a trench in the ground before the excavation. By adopting these measures, the wall displacement and bending moment were kept within the design value and the excavation was safely completed.

Table 1 presents the results of a survey at 100 construction sites showing the behaviour of earth-retaining walls during excavation. The structures under construction were building basements, shafts, sewage treatment plants, subways, underground tanks, etc. Each excavation depth was more than 20 m and the maximum was 46 m. Most of the retaining walls adopted were reinforced concrete diaphragm walls or continuous pillar walls. Most of the designs were done by framed structure analysis or simple beam analysis.

According to the survey, underestimate or overestimate of the wall displacement was observed in 43 sites and that of the wall stress in 48 sites. Over-design sometimes results in uneconomical construction, and an underestimate of wall stress and displacement is very dangerous particularly in deep and wide open

TABLE 1 *Comparison between design and practice (no. of sites).*

Comparison of the observed value to the design value	very small	smaller	nearly same	larger
Displacement of wall	29	34	23	14
Stress of wall	41	36	16	7
Axial force in strut	29	57	10	4
Settlement of back ground surface	40	33	18	9

excavation. The gap between design and practice was probably caused by an under- or overestimate of the soil pressure.

These experiences taught me the following lessons:
- Design values always include uncertain assumptions.
- The important point here is not the magnitude and distribution of soil pressure but the magnitude of the wall displacement and the axial force in the struts.
- Monitoring is the best way to understand the behaviour of a real structure.
- Safety is paramount in construction. Therefore, prediction of structural behaviour before action is important.
- Preparation of countermeasures is also important.

The method of prediction presented here is still used widely. Each construction site can also be considered as a very good practical experiment. The back analysed soil pressure is useful for design of retaining walls in similar ground conditions. Accumulation of such data at many sites will lead to more rational design of earth-retaining walls in open excavations.

Cut slope stability

Most of the land in Japan is mountainous and cut slope works are, therefore, very common. However, it is, in general, not easy to evaluate cut slope stability. As technical advisor I was in charge of the construction of a power station. The key part of the work was to reduce a very unstable mountain up to 33 m in height with the slope gradient of 1 to 0.3. In order to avoid the slope sliding or application of unnecessary reinforcement to the cut slope, it was essential to evaluate the slope stability properly. In this construction, I developed a prediction method for cut slope stability. A brief explanation of the method follows.

A rough idea of the soil or rock profile of the mountain and the approximate value of the elastic modulus in each layer can be obtained by inspection and testing of bored samples. The elastic modulus E obtained by the uniaxial compression test of bored samples does not accurately represent the elastic modulus of the layer, since the sample is taken from the relatively sound part of the layer even if the layer includes many cracked zones. Furthermore, the elastic modulus will decrease when deformation of the layer increases due to the excavation of the rock. It is, therefore, not easy to calculate the behaviour of the mountain. Monitoring is employed here again.

Figure 2 shows a cut slope and measured displacement of the mountain by cutting. The displacement can be measured by inclinometer through the borehole. The potential sliding surface is assumed to be the straight line connecting the point where the abrupt change of the displacement occurs and the toe of the present cut slope. If the displacement data from several boreholes are available, the potential sliding surface can be obtained as a curved line. Assuming the elastic

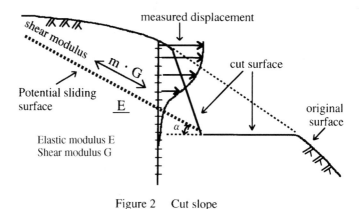

Figure 2   Cut slope

FIGURE 2 *Cut slope.*

modulus of the mountain to be E and reducing the shear modulus of the layer above the potential sliding surface, as $m \cdot G$ in the direction of the potential sliding surface, G being the shear modulus of the mountain and m being less than 1, the displacement of the mountain can be calculated by the finite element method, applying a force evaluated from the stresses applied on the present cut surface before the cutting. In this calculation, the Poisson's ratio is set to a standard value if adequate data are not available, and is assumed to be unchanged. E and m are determined so that the calculated displacement matches the measured ones. If a mountain consists of several layers, an individual value of E is assumed for each layer. If the displacement is evaluated properly, the calculated strain is considered as a good estimate of the real strain.

The present slope stability is judged by comparing the maximum shear strain ε to a control strain εc, such that the safety factor is $S_f = \varepsilon c / \varepsilon$. εc is selected, for example, as the ratio of the compressive strength to the initial elastic modulus of the bored sample, limiting the strain of the mountain within the elastic range.

By this method, the safety factor of the slope after the next cutting can be evaluated before cutting. If the stability is found to be insufficient, proper action can be taken before the next cutting. The power station was constructed safely although rock anchor reinforcement to the slope was necessary at certain levels of the work.

Adoption of the prediction method of cut slope stability made the construction period shorter and the construction cost lower. Furthermore, the construction workers were able to carry out potentially dangerous work without anxiety. This method can be applied to monitor the stability of many different types of slope.

A lesson I learnt during this project is that better understanding of construction can result in reasonable construction costs together with better safety assurance.

140

Shield tunnel

In a major project involving a shield tunnel, 15 m in diameter and 9.5 km long,across the Tokyo bay it was extremely important that it be accurately designed, not only from the safety point of view but also for economy.

Tremendous effort was put into the design. Various types of design models for the tunnel were discussed. Spring constants representing the ground surrounding the tunnel were required, together with the soil and water pressure acting on the tunnel and the rigidity of the tunnel itself. Moment distribution in the tunnel taking into account the effect of segment joints was also very important.

The spring constant could be evaluated from the modulus of subgrade reaction, K. It was, however, uncertain how to estimate the K value for such a large diameter tunnel from the value obtained in a very small diameter borehole. At that time, I was involved in the construction of a shield tunnel under the seabed. The diameter of the tunnel was about 5.5 m and the length of the tunnel segments was 1.0 m. I proposed a model test of the tunnel and loading test of the real tunnel.

The model tunnel consisted of five rings assembled in the laboratory with the same segment as used in the construction. Uniform load was applied to the model by hoop prestressing corresponding to the soil and water pressure. A line load was also applied to the tunnel from the inside by several jacks along the tunnel length through steel beams. From this test, the rigidity of the tunnel was evaluated and the maximum bending moment of the model tunnel was found to be in the order of 20% higher than the average value.

After completing the model test, the loading test was performed in the real tunnel. A line load was applied to the tunnel from the inside by five jacks through steel beams along the 10 m length of the tunnel, creating a plane strain state at the middle of the loaded span. The load was applied in both the horizontal and vertical directions, and the deformation of the tunnel measured.

Calculation of the tunnel deformation caused by the same loading as the test was also performed assuming various spring constants for the surrounding soil. Then, the best fit of spring constant for the soil was selected so that the calculated deformation matched well to the measured one. The modulus of subgrade reaction to the larger tunnel diameter was evaluated from the selected spring constant.

It was very interesting that the ratio of the back analysed K value for the larger diameter D to that of the borehole diameter D_0 was nearly $(D/D_0)^{-3/4}$. Measurement of the axial force and bending moment occurring in the tunnel also made it possible to back analyse the soil and water pressure acting on the tunnel. These results were taken into consideration in the design of the Tokyo bay tunnel.

The lesson I learned here is that we should not miss a chance in any construction work to perform useful experiments.

Earthquake

In 1995, Japan suffered enormous damage from the Great Hanshin-Awaji earthquake, of which details have been published elsewhere. This earthquake taught us many engineering lessons, from which I, too, have benefited. In respect of reinforced concrete bridge piers, this earthquake revealed many deficiencies of modern design technology, as various types of failure were observed in bridge piers, some of which had never been anticipated in design.

For instance, thousands of people living in the Kobe area felt a very strong upwards shock right at the beginning of the earthquake, and many phenomena which might be caused by the upward shock wave were reported. Most civil engineers, however, ignored this fact, probably because they did not experience the earthquake themselves and the upward wave was not recorded by seismometer. This shock wave might have initiated successive failure in many structures. If so, this effect should be included in design, and structures should be properly reinforced against a future earthquake. A lesson I learned is that we should not ignore what happened and that we should accept it in a humble manner without preconception. Many houses and buildings were seriously damaged or destroyed for reasons such as top-heaviness, lack of bracing, etc. There are many similar houses in Japan, and most of the people living in them admit to a fear of earthquakes after the disaster. However, most remain unmodified as yet against future earthquakes.

A lesson here is that we should not repeat the same mistake.

Although I would like to mention many other experiences, I will close this essay with a few more personal opinions.

Due to the vast development of our modern world, each area of science and engineering has been broken down repeatedly into new specialisms. As a result, countless specialities have arisen. Take a look just at civil engineering. There are so many areas and sometimes we do not know what research is going on next door. There is an undesirable tendency in many Japanese universities for most graduate students to study only a limited field of their choice, probably due to too much concern about their graduation thesis. This tendency is often apparent in the research centre and even in the design department of many construction companies. There is another undesirable tendency, in my opinion, for some Japanese students to prefer to do their research only by computer rather than through experimental work using their own hands. I am not just criticising the use of computers. I am saying that we can learn a great deal through experiments and experience at real sites. I would like to send the following messages to young engineers.

- Learn more from failure than success and do not repeat the same mistakes.
- Do not over-rely on computers but look more at real phenomena.
- Do not be conservative but progressive.

- Do not be a specialist but try to be an all-round engineer.
- Visit various construction sites, where you can learn a lot and gain new ideas for a future challenge.

Construction sites are the best teachers of construction engineering.

Professor Koicho Ono

with my wife Masako

BSc, MSc, PhD

Born in Osaka, Japan. After graduating from Kyoto University he became a research assistant at the University of Calgary and in 1972 obtained his doctorate from the University of Toronto. After over 20 years in the construction industry with Konoike Construction Co. Ltd (and as visiting lecturer at Kobe University) in 1996 he was appointed Professor of Civil Engineering at Kyoto University.

He has worked on numerous committees and undertaken international activities, including: fatigue design for RC slab, JSCE (1975–78); design of deep foundation, JHPC (1977–79); technical committee for concrete structures, HHPC (1979); estimation of safety for bridges, JSCE (1983–85); alkali aggregate reaction, JSCE and JCI (1984–86); construction and durability of concrete structures, JSCE (1988–); shotcrete WG, Vice Animature (since 1995), ITA (1989–); General Secretary of Canada-Japan workshop on advanced composite material for concrete structures (1993–95); General Secretary of South East Asian symposium on tunnelling and underground space development (1994–95); International conference on shotcrete, Chairman for Asia (1995); large scale excavation, Ministry of Construction (1995–96); Kobe earthquake, JSCE (1995–98); Vice Chairman, the Second International Congress on Environmental Geotechnics (1996); rock support in tunnelling, JTA (1996); shotcrete improvement, JRCC (1997); the Second International Conference on Analysis of Discontinuous Deformation (1997); Asian natural gas pipeline (1997); General Secretary, seventh Korean, Taiwan and Japan trilateral seminar/workshop on civil engineering (1997).

He is also International Advisor to the JTA (1997–), Technical Advisor to the HHPC (1998–) and Director of the JCI (1998–).

JSCE	Japan Society for Civil Engineering;	JHPC	Japan Highway Public Corporation;
HHPC	Hanshin Highway Public Corporation;	JCI	Japan Concrete Institute;
ITA	International Tunnelling Association;	JTA	Japan Tunnelling Association;
JRCC	Japan Railway Construction Corporation.		

Civilization through civil and structural engineering

Milcho Brainov

Summary

All civil/structural engineering works (buildings, bridges, etc., in general, all material things with definite form in space) are first of all structures, in particular, load-bearing structures. Civil/structural engineering has been of fundamental significance for the civilized world, and will be in the future, as a complex of science, practice and art. Civil/structural engineering errors – as a result of lack of knowledge, competency, legislation and control – are inadmissible. The consequences are fatal: destroyed buildings and structures, human victims and economic damage.

Civil/structural engineering is not only or simply calculation and technologies, but something more and greater. This essay aims to present in a systematic and orderly way the reasons why structural engineering is 'more and greater' and what is its significance and role for human civilization until today, on the eve of the third millennium. At the end of the essay, particularly, I shall point out what encouraged me at the time to enter this great profession with its endless fascination, challenge and intellectual reward, and why my professional career induced me to write this essay.

Introduction

The well-known human and international topics of the built and economic environment, natural disasters reduction, sustainable development, etc. have come to be permanent for the civilized world. All these problems are connected with artificial (made by people) changes to nature and especially the material changes brought by building and all other civil and structural engineering works in the widest sense. They are the fundamental conditions for the functioning and development of societies, states and the global world. They are also the basic conditions for production and activities related to food and other goods, standards and quality of life, health and social care, culture and amusements, cults and rituals, etc.

In the contemporary civilized world all these are impossible without civil/structural engineering activities

- on – all aspects of planning, design, construction, operation, inspection and monitoring, maintenance and repair of structures, regardless of construction material, type of structure or construction method.
- of – civil/structural engineers in cooperation with owners, architects, engineers (mechanical, electrical, fire protection, foundation.), consultants (wind, vertical transportation, security, plumbing, automation, acoustical, code, etc.), with contractors, fabricators, erectors, developers, rental agents, project managers.
- from – engineering offices, construction firms, administration and civil services, universities, institutes of technology, research institutes.

Civilization

Human civilization is a product and expression of technical (material) and political (social) changes in nature in two distinctive aspects: social and material.

Social civilization is an expression of social (political) development. From 'natural people' living on a 'natural planet' humanity long ago moved into 'civilized societies'. At different places, in different times, with different pacing and at different levels, societies have passed and still go on through well-known different social historical formations. Whether aware of it or not, human societies are striving towards a world association and global integration.

Material civilization is an expression of material (technical) development. Human societies motivate their social civilization on the basis of qualitative and quantitative civilized material-technical changes of the natural planet. Such are the buildings and structures in a broad sense.

  Buildings – all types: residential/housing, administrative, sports, hospital, educational, hotel, cultural, industrial (light, heavy, etc.), agricultural (various types), as well as all other territorial, town planning and architectural projects.
  Structures – all types: bridges (highway, railway, etc.), towers and masts, reservoirs and gas holders, technological (chemical, metallurgical, etc.), power stations (thermal, water, nuclear, etc.) transmission lines (electrical, gas and others), dams, sluices, canals, tunnels, harbours, quays, bus and railway stations, airports, technical infrastructures (water supply, heating systems, etc.) as well as all other civil/structural engineering works.

These are all artificial (created by people) environments and systems for the life and work of the people. So human civilization is impossible without these artificial environments and systems, which are products first of all of civil/structural engineering activities.

Civil engineering activity

The big difference between a natural and a civilized planet – between nature and civilization – can be illustrated not only in social, but also in material aspects. The widely-used trinity man-society-nature contains the natural 'man' (natural people) and his development in society as a social civilization. But nature (the natural planet) is left unchanged without being materially civilized. Therefore, the trinity man-society-nature has to be replaced by the global unity 'nature-civilization': where nature shall have the meaning of natural people and natural planet, while civilization means not only social civilization, but also material civilization. In this respect civil/structural engineering plays a principal and fundamental role.

Normally, people speak about civilization in general, associate it too much with 'art', less with social civilization, and almost never with civil engineering works. Even impressive buildings, bridges, etc. are considered as 'works of art', but not as achievements of civil (architectural and structural) activities.

Civil/structural design

Every building, bridge or any other civil/structural engineering work originates from some function (purpose) which in its turn requires an appropriate form. The forms are actualized as a structure, according to an existent technology. Civil/structural creativity is entirely subordinated to the full unity between function-form-structure-technology. Here we meet the dialectical contradiction between function and form as a necessity, and structure and technology as a possibility.

The civil/structural engineering design is the unique professional creative activity, which simultaneously renders an account of (and is answerable for) all function-form-structuretechnology.

Criteria

Criteria are necessary for assessment of the optimality or expediency of any civil/structural decision. They are:
firstly, qualitative criteria, for a complex maximum of
  • functionality/serviceability
  • durability/sustainability
  • reliability/safety
  • aesthetics/beauty, etc.
Secondly, quantitative criteria, for a complex minimum of
  • material
  • labour
  • energy
  • time
  • price, etc.

Thirdly, conjunctive criteria, observed in
- regional economics
- technical culture
- degree of local development
- experience gained by civil/structural engineers, technicians, workers, etc.

Structural design creativity

The structural design process is a specially creative one , in three methodological points. The first is based on the kinds of creative thinking:
- abstractive
- extrapolative
- associative
- foresighting (considering the consequences).

The second point is based on creative methods:
- differential analysis
- integral synthesis
- multiplying method
- deduction method

The third point is based on creative approaches:
- systematic
- complex
- programmed
- aiming.

Creative structural design does not tolerate routines. Therefore, it cannot be automated and computerized as a whole, but only as a computer aided design (CAD) within the responsibility of the structural engineer, the author of the project or design.

Structural creative design is achieved by a creative design approach. It determines the possibilities: (1) in variants, of many alternatives; (2) in expedience, with all dependencies; (3) in statics, as states; (4) in dynamics, as a process; (5) in time, as a chronology, etc. Only by this design approach can subjectivity, conservatism, incompetence, etc., whether personal or of a group, be avoided. Structural designers can imagine that the consequences of these would be catastrophic!

Civil/structural requirements

The structural design is subordinated also to fundamental requirements about:
- economy and efficiency
- safety and reliability
- serviceability and durability
- aesthetics and beauty

By including functionality, durability and beauty with efficiency as a degree of standard, and by laying down terms as guaranteed priorities, it is possible to obtain two complex fundamental requirements: 'economy and efficiency' and 'safety and reliability'. These are, by the way, the eternal fundamental requirements in everything. Both are within the competence of civil/structural engineers. But the second requirement – safety and reliability – is within the competence and responsibility of structural engineers only, and of nobody else.

Civil/structural investment

When talking about buildings and structures, everybody must be aware that their erection and development needs investment, implemented in four basic investment stages – planning, design, construction and operation – and that: (1) the optimal/expedient solutions are to be achieved during the design stage; (2) the application of the results of technical progress is possible only through design; (3) their skilful execution at the construction stage must be well provided for by the design; and (4) structures must fulfil their functions for a long period of operational time, which must also be well specified by the design. The above do not reduce the significance of the other civil/structural investment stages.

Underestimation

Civil/structural engineering must give special attention to avoiding underestimation of the following: (1) the design which affects negatively the social prosperity, the contemporary aspects of the technical progress, the expected technical economic effect, the functionality, serviceability, safety, the prevention of incidents, disasters, the loss of materials, the crucial problems of socio-economic planning as well; (2) the operation which may cause huge economic losses, human victims, reduction of 'physical and moral life' of the buildings and structures in general, their safety, etc.; (3) the construction which must be developed in accordance with the design solutions, but not to the opposite extreme, as a barrier to future progress.

One comparison

Structural design can be compared to one of the most impressive arts, music. The 'products of music' are composed (designed) by composers (designers) and are performed (executed) by performers/executors. Musical composition is very similar to structural design in another way too. Both musical composition and structural design represent creativity based on much science and theory: the musical composition on the science and theory of harmony, counterpoint, solfège, rhythm, etc.; the structural design on the science and theory of structural shaping and modelling, loads and impacts, materials and compositions, structural mechanics and behaviour, dimensioning and detail, etc. Structural design (similar to architectural design) is qualitatively different from the design of industrial products. Industrial products are mobile, produced by stationary production capacities on many individual sites; the civil/structural products on the contrary are stationary, erected by mobile capacities on individual sites.

Besides, structural designs do not have preliminary prototypes, zero series, licences, etc., of the kind known in industry, due to high costs, large sizes, slow erection, and the underlying uniqueness of buildings and structures. Therefore, structural design is based on scientific forecasting, abstract thinking and creative imagination, with no right to admit errors. The consequences of errors in respect of the functionality, reliability, durability, aesthetics, material, labour, time, finance and energy losses, are irrevocably fatal. As far as structural engineers are concerned they will be personally responsible for failures in safety.

Further, structural design necessitates independence and creative individuality. This is absolutely valid for bridges and many other structures, where the individual role of structural engineers is indisputable. This is also partly valid for buildings, where the role of architects and structural engineers is mutually integrated. For example, note how many buildings and structures with architectural image and aesthetic impression are the result of the structural engineer's role. Historically, architecture and structural engineering have been united. Today, as professionals and specialists, as activity and creativity, they are divided, or rather differentiated. But in the completely natural division of their work in a contemporary, highly organized, civilized world, they are always functionally united, professionally integrated.

Science–technics–culture–art

Finally I would like to see structural engineering in harmony with science–technics–culture–art. At the time of Leonardo da Vinci's anniversary in 1980, I stressed in one of my articles 'the necessity of increasing the general and artistic culture of civil/structural engineers (scientific-technical intellectuals, in general) as a mighty factor for provoking and developing their scientific-technical creativity'. And further, 'in this respect, but not in a lesser degree, the artistic and cultural intellectuals should increase their general engineering and scientific-technical knowledge as important factors for their representation and motivation of the civilized world (both in the material and social sense)'.

In this respect I think that the scientific-technical (engineering) intellectuals have travelled a long way. But the fine art intellectuals are falling behind. And worse, instead of trying to increase their scientific-technical knowledge, they often prefer to denounce it.

The harmonious unity between science–technics–culture–art is first of all a challenge to civil/structural engineering intellectuals, which take them for guarantors in respect of the survival and development of human civilization. I have in mind primarily the material civilization for which civil/structural engineers are objectively and directly responsible.

Conclusion

The above philosophy of civil/structural engineering expounded for the purposes of this essay would be sufficient, I hope, to appraise the role of civil/structural engineers and their significance for society and the global world. The ideas put forward are intended to be of some help to civil/structural engineers in respect of the quality of their professional activities, and on this basis, to raise their prestige in societies and states in a global world. These ideas demonstrate also that civil/structural engineering is really a creative activity, demanding competent personalities and specific talents.

My thoughts in this essay are the result of knowledge and experience gained in my civil/structural engineering professional life and career. These thoughts, knowledge and views might be the reason for my broadly-based and interesting career with its endless fascination, challenge and intellectual rewards.

Professor Dr. Milcho Brainov

Diploma in civil/structural engineering in 1950 at the State Polytechnic, Sofia. Doctor of Technical Sciences and then Professor of civil/structural engineering. University activities (1950–90): Assistant Professor (1950), Associate Professor (1958), Professor, Head of Structural Department (1965–90), at the University of Architecture and Construction, Sofia, Bulgaria.

Scientific/research activities: structural mechanics, civil/structural (incl. bridges) engineering. Author of more than 100 works, including books, monographs, papers, articles, research works, manuals; he has been Chairman of the Structural Engineering Council of the State Higher Testimonial Commission.

Civil/structural engineering activities: construction of the Danube Bridge, Romania-Bulgaria; over 50 civil/structural design projects (the biggest in the country: Asparuch Bridge, Varna; National Culture/Congress Centre, Sofia); consultant in many technical projects; President of Metal Structures a state company.

State/public activities: in parallel with the above activities, he has been Head of Department at the State Committee of Construction and Architecture; Member of the National Delegations in International Committees for structural and construction codes; Deputy Minister of Higher Education; Vice-President of State Committee of Culture – on cultural investments; Councillor of the Council of Ministers – on Science and Higher Education; Member of ExC of Union of Scientists in Bulgaria; Founder, President and Honorary President of the National Union of Civil/Structural Engineers.

International activities: Vice-President and member of the ExC (1986-91) of the International Association for Bridge and Structural Engineering (IABSE); Representative (1976–) of the Council on Tall Buildings and Urban Habitat (CTBUH); Liaison person and Focal Point (1990–96) of the International Decade for Natural Disasters Reduction (IDNDR). Visit to numerous countries to participate in forums and meetings.

The background of good design abilities

Fritz Leonhardt

Civil engineering is a fine profession and can give deep satisfaction if the engineer has learned the art to design in a broad way. To make good designs is an art and, like any art, requires intensive studies and work to become a master.

The attitude towards this art of engineering must be that of serving our fellow citizens. Engineers and architects design buildings, structures, traffic ways, bridges, harbours – all needed to serve society, to provide living and homes for masses of people, to improve their living conditions, not only materially but also in culture and fine arts. Therefore we must be aware of being servants and should avoid serving personal vanity and prestige. We should also be aware that the two professions of architect and engineer belong closely together and should cooperate whenever they have to design buildings and structures.

Judging and analysing structures

Doing this in a positive way, both must be aware of the many different possibilities of choice. The young engineer needs to see many structures and register them mentally. First he must learn to see consciously, to observe and analyse the qualities of the building or structure and to consider the effects on human beings as users or neighbours. In order to analyse a structure, one must have learned the basics of structures. This means the engineer should know the structural systems like beam, slab, plate, frame, arch, suspensions, folded plates, shells, trusses, space trusses. He should know their reaction and resistance to acting forces such as dead load, live load, wind, temperature effects, impact, etc. Judging structures requires a basic knowledge of mechanics, statics, strength of building materials and soil mechanics. He should also have made or seen tests on structural members up to failure.

Editor's note
This essay is an edited version of a paper presented by Professor Leonhardt at Bath University in 1984. It is presented by courtesy of the International Association of Bridge & Structural Engineering.

These fields concern mainly stability and safety against failure, which must be guaranteed in any case. But the effects of a building on human beings are equally as important as safety. We have to design for the comfort of the user, for economy for the owner in regard to operation, energy consumption, maintenance and durability. To judge these qualities, one must have basic knowledge in fields such as building physics concerning thermal effects, moisture, noise, acoustics, light and colour. Some knowledge of the physiology and psychology of man is needed to understand the human response to such qualities. Cost calculations for investment, operation, maintenance, etc., should be understood. For durability, the behaviour of building materials under climatic and chemical attacks must be known. All these sciences require fundamental knowledge of mathematics, physics and other basic natural sciences, as are usually offered at college level.

The student's problem: basic knowledge in many fields

For a student it is almost a deterrent to become aware of this wide range of knowledge which should be available for good design ability. But please recall that I repeatedly spoke of basic knowledge, knowledge of the principle laws of performance, sufficient to choose the right solution in principle. One can always call on a specialist for the detailing. In practice, there is usually a team of specialists for the design of any important building, but it is always helpful if the members of such a team have some basic knowledge in the different specialized fields, which helps them understand the specialist and prevents the danger of one-sided partial dictates of specialists.

Studying architecture or engineering therefore means learning the basics of many sciences, to have them as tools for designing. These sciences shall not govern but serve and help to fulfil the purpose, to satisfy functional needs and requirements. These functional requirements have a great variety. Sullivan's famous slogan 'form follows function' was never meant to refer to structural form only, but to the whole building, which must function not only in physical requirements but also for psychological and emotional desires to provide wellbeing and harmony. This involves aesthetic qualities, as is clearly stated by Fraser Reekie in his book *Design in the Built Environment*,[1] but is not implemented in practice.

Grasping practical experience

Learning acquired in universities or other schools is not sufficient for becoming a good engineer. The student should travel a lot and see and analyse many buildings or structures. He should make sketches or take photographs. To be skilled in free-

---

[1] Reekie, R.F. *Design in the Built Environment*, Bristol Polytechnic, ...

hand sketching is most important, and often neglected nowadays. He should visit many construction sites and workshops of craftsmen or fabricators or, even better, work there for some while, because he needs to know how materials are machined or made and put together or how construction or erection is carried out. He should have construction and fabrication in mind when he designs larger structures, because quite often construction sets limits or influences costs and economy.

The student should also not hesitate to contact the engineers who are responsible for the structure which he visits, or the field and resident engineers on the site, and ask them the many questions which come to mind when trying to analyse what he has seen. Such field discussions can be most helpful. When studying buildings he should also question the users; he would often be surprised to hear the complaints by users about poor performance of buildings, even those designed by famous architects.

This background of learning is most important, because the design abilities improve with the quantity and quality of structures which the engineer has consciously seen, studied and registered mentally. He must have a rich repertoire of design possibilities to draw from when he starts designing for himself.

The design process

Design begins with collecting data concerning functional requirements, the use and purpose of the building, and the desired visual effect. Further data are needed about the site, ground conditions, climatic situation and orientation, economic limits, neighbouring and environmental consideration, local availability of materials, power, skilled labour, etc.

With these data in mind, the structure gets its first hazy shape in the imagination of the designer. The pencil transfers the concept to paper. After the first sketch a second, a third sketch is made. Different solutions are compared, evaluated and put aside for further brooding and contemplating in the mind, best overnight. Then new sketches are made, now to scale, checking the various and manifold requirements, looking at proportions of building masses, etc. At this stage, the designer, architect or engineer should show his sketches to his partners. The task of designing is so complex in our day that nobody should imagine creating a design all on his or her own. Early discussions with experienced or specialized partners are most helpful because three or four heads always see and think better than one and they look and judge from different viewpoints. This early collaboration is mainly necessary between architect and engineers. Criticism should be welcomed and digested. In this way the design begins to grow out of infancy. It is still a long way to maturity and implementation.

If one thinks that the design will fulfil all functional requirements, then it should be shown to and discussed with laymen, with friends, preferably with ladies of good taste. I do this even with my bridge designs, and it is often surprising how their impression and remarks help to improve the design.

Design analysis and the role of the computer

So far no statical calculation has proven necessary because the required structural dimensions are usually roughly known by experience. But for the next step of the design procedure, rough hand calculations should be made in order to get the dimensions of girders or critical structural members sufficiently large for sound constructability. Some structural detailing may also be done early for critical points.

Many engineers these days run large computer calculations before they have a clear picture of the structure and its important details. The computer should not be used before the structure's design has reached some maturity. The computer is undoubtedly an important aid for the engineer, especially in the final state, but it is often misused to impress the client and to waste a lot of paper. The ability to make rough hand calculations must be cultivated for the future, mainly for quick and reliable checking of computer output and for developing sound judgement of structural dimensions.

The engineer must also be aware that the computer cannot design and cannot produce innovation. Creativity is still the privilege of the human mind, and creativity is badly needed to solve the design problems of the future. Therefore we must place the computer into its correct role, as a technical tool. With good software, it is indeed a very capable aid, allowing complicated calculations which were not possible in former times. As an example I may mention the Olympic Stadium in Munich. The calculation of the exact geometric lengths of the many thousand strands and ropes for the networks could only be done successfully by computer, for which Professor John Argyris prepared a special program just in time.

Most design can be based on well-proven and established solutions out of our repertoire, but the tasks and means change and therefore we should always search for improvements compared with the traditional approach. This means critical thinking: to be doubtful, sceptical and suspicious about the soundness of earlier solutions. The human mind is not perfect, and a lot of our views and thinking are based on traditional prejudices, sometimes on superstition or on the effect of the famous hidden persuaders (Vance Packard) which are engaged by the salesmen of our big industries. This is especially true in architecture which is so much governed by blatant fashions. Just take the façades of buildings, especially those of high-rise buildings.

Some functional aspects

Essential functional requirements are not fulfilled if such buildings have glass façades in the heat of Texas as well as in Calgary with temperatures of $-30°C$, and if identical curtain walls face south, east, west or north. The reflection of clouds on the glass may please an architect as it is reported of Mies van der Rohe, but it does not benefit the user who has to suffer behind the glass or the owner who has to pay the bill for energy consumption in summer and winter time. Large areas of façade covered with plain concrete in the style of brutalism cannot satisfy either; they injure our natural aesthetic senses.

The exposure of the structural skeletons of multistorey buildings or even the exposure of installation pipes (Centre Pompidou, Paris) is nothing else but a search for sensation to satisfy vanity and has nothing to do with fulfilling functional requirements. God has protected the skeletons and bones of all his creatures, including man, with a thick pelt and skin. So our buildings need protection against different climatic attacks, and this protection should be different, depending on the type and strength of these attacks.

Architecture in crisis

The architecture of the last 50 years has reached a serious crisis, mainly because it did not fulfil the basic human requirements of comfort, well-being and social needs. The needs of the psyche were almost totally neglected. The sense of aesthetics and the capability of sound aesthetic judgement has been widely lost in our whole society (not only amongst architects). Ugliness of the built environment makes the human soul sick and causes depression or aggression. In an ugly environment crime grows like mushrooms. People long for more beauty, for more humanitarianism in our cities.

Architecture should not be styling building exteriors to follow fashions like functionalism, international style, brutalism, post-modernism and such ideologies. Architecture must develop to fulfil human, physical, psychic and emotional requirements. This is a challenge in the near future for creativeness in our professions. Engineers and architects must relearn the basics of psychic desires and requirements, the basics of aesthetics and also the basics of all that is needed to make users feel comfortable. Ecological requirements also pose a challenge – a wide field from which I mention only the necessity to get clean air and water, to save energy and to switch to solar energy.

The role of building physics

A new architecture satisfying these requirements cannot be gained by nostalgia, by returning to Palladio or other old styles. We must get away from considering architecture mainly as styling, as shaping for exterior visual appearance. The new façades must be developed to function in all the aspects which I have mentioned: better thermal insulation, sun shades different for west, south and east which give thermal insulation to windows at night in winter time. We need new heating systems such as electric low temperature radiation from ceilings or walls which in combination with more thermal insulation allow energy consumption to be reduced to 30% of what we use now. In addition, such heating avoids air pollution. We also need good natural ventilation of rooms, better acoustics, etc. There are so many things which must be changed and improved in order to gain satisfactory living conditions in our cities.

This challenge mainly concerns engineers who, by their training in rational thinking, by their scientific approach, technological knowhow and creativeness, can help find new solutions for a better architecture. For this, knowledge in building physics is more important and fruitful than mechanics and statical analysis.

All this is a great challenge for the next generation – learning in a broad sense, learning also outside the universities which often are limited to old and well-established sciences. It must be a learning from real life, from the needs of our fellow men. The challenge calls for creativity, for research and innovation, not in opposition to technology, but towards a more human technology that also considers ecological aspects. We can have confidence in the future, but we must get to work with courage in order to solve the problems.

Importance of aesthetics

I have mentioned aesthetics several times. For me there is no doubt that the aesthetic qualities of our built environment are as important as the economic materialistic aspects which have had priority so far. Aesthetic qualities have an enormous influence on social and ethical behaviour and thereby on the psychic health of man. Buildings and structures have aesthetic qualities which impress people, mainly subconsciously. Man has a natural sensitivity for aesthetic qualities in the visual appearance of his environment. The majority of the people of a certain cultural origin agree on the judgement if objects are classified as beautiful or ugly. Beauty causes pleasant feelings and can be enjoyed; ugliness causes discomfort and can even hurt. P.F. Smith[2] says: 'aesthetic perception has developed into one of the highest capabilities of our nervous system and is a source of deep satisfaction and joy'. We must relearn to analyse these aesthetic qualities, and we shall find that there are rules for good and bad in aesthetics and that the old master builders have applied such rules with great success. Such rules or guidelines are outlined and explained in [3].

Aesthetics is a wide field for learning and this learning should be done by our whole society in order to gain more understanding of the real values of aesthetic qualities for human satisfaction, values on which it is worthwhile spending money. Such learning is also needed for better designing abilities, which will give satisfaction and happiness by better design results. Learning pays!

R. Fraser Reekie speaks in the preface of his book of 'the urgent task of abolishing ugliness, dreariness, squalor and all offensiveness from towns, villages and countryside, restoring and producing visual pleasure in the environment, so that life can be lived therein more healthily and happily'.

[2] Smith, P.F., *Architecture and the Human Dimension*. London, 1979.
[3] Leonhardt, F., *Bridges, Aesthetics and Design*. DVA, Stuttgart, 1982.

Closing remarks

Remember, it is not only the visual appearance of our built environment from the outside, but also the comfort and beauty of the inside of our buildings, where we spend much more time, which counts for a healthy and socially sound development of the future of mankind. Engineers and architects are challenged to learn, search and work for this aim.

Manifold studies in different fields and an awareness of human requirements are needed to acquire good design abilities. Cooperation with specialists can improve the result. With this background, designing is a pleasure and efforts are rewarded by satisfaction. Most importantly, designing should always be done with the attitude of serving and giving, which brings more satisfaction than pretence and taking.

Fritz Leonhardt

Prof. em. Dr.-Ing. Dr.-Ing. E.H. mult.

Born in 1909, Fritz Leonhardt became a student of civil engineering at his home University in Stuttgart. After graduate studies at Purdue University in West Lafayette, Indiana, he returned to Germany as a bridge engineer. In 1939 he founded Leonhardt, Andrä und Partner GmbH, Stuttgart, consulting engineers. From 1957–1974 he was Professor of Stuttgart University (later Emeritus) and between 1966 and 1969 was also President of the University.

Fritz Leonhardt has been involved in many developments within civil engineering notably bridges, prestressed concrete, and crack control. His work on bridges has included prestressed concrete railroad bridges, development of new systems for suspension and cable stayed bridges and steel and concrete bridges for high-speed railways.

He is the author of many scientific papers, research reports and a number of books on prestressed concrete, bridges and towers.

He has received innumerable awards and honours, both national and international among which are: Werner-von-Siemens Ring (1965) the highest German honour in the field of technical achievement; Freyssinet Medal of the FIP (1974); Gold Medal of the Institution of Structural Engineers, London (1975); Award for Engineering Excellence from the American Consulting Engineers Council (1979); Grant Order of Merit of the Federal Republic of Germany (1980); Foreign Associate of the National Academy of Engineering, USA (1983); Prix Albert Caquot 1989 de l'Association Francaise pour la construction; School named after him in Stuttgart-Degerloch (1998). In addition he has received honorary degrees from the Universities of Braunschweig, Lingby in Denmark, Liège in Belgium, Bath in England and Pavia in Italy.

The changing face of engineering

Tom Smith

No country in the world has produced better engineers or influenced the science and art of engineering more than the United Kingdom. Why is it that, with our history, we are no different from all the other nations of the world in that we are now failing to attract into engineering the brightest of our young people? Why is it that, of those who graduate as engineers, only one-third have entered the engineering professions, the majority choosing accountancy, insurance, the money markets and other 'uninteresting' professions?

Something is wrong; is it image, is it reward or is it simply that communication between we practising engineers and you young aspirants is so poor that we fail to transmit the excitement, the fun and the satisfaction in achievement which comes from being an engineer? I look back on forty years in engineering at every level and I ask myself what it gave me and what it offers you, our possible successors.

There are so many answers to the former question that I must limit myself to only the major benefits: travel throughout the world to both pleasant and hostile areas, friendships formed with interesting and stimulating people in many fields of work, opportunity to tackle and, in most cases, solve problems which have stretched me physically, mentally and intellectually and in the process acquiring a great sense of achievement in helping my fellow man towards a better life. Financial rewards came in sufficient quantity to be more than adequate for my needs and now, in the autumn of my engineering life, the knowledge that I would not change my role of 'engineer' for anything else is probably the most selfish satisfaction of all!

What does it offer you? I must again restrict myself to areas of personal experience as a consulting engineer. I am myself a building services engineer but my interests as a past chairman of the Association of Consulting Engineers range widely over all disciplines of engineering. I feel that opportunities for engineers in the future are best illustrated by the contribution they can make to urban renewal throughout the world and, in this respect, the startling and unprecedented political, economic and social changes occurring everywhere presage the most exciting decade you are likely to experience in your lifetime.

Engineers have a vital role to play alongside government departments, developers, financial institutions, and contractors in teams promoting individual development schemes for commercial, residential, leisure, industrial, health and many other purposes. As well as engineering expertise, the engineer will bring to such projects cost control and project management skills to ensure that construction is achieved to the quality required, in the time required, and within the budget available.

An enormous contribution will be made by engineers towards improving the rapidly deteriorating quality of environment we are experiencing throughout the country and indeed worldwide. It is their responsibility to assess existing contamination sources and projects which hazard the environment and to propose methods and schemes to correct or remove the risk. They design waste and effluent systems to enable control of polluting processes and in the development of new projects. They make a thorough investigation of the environmental impact and the prime measures to obtain the most benign environmental conditions. In building services they apply skills to minimize the output of carbon dioxide into the atmosphere from heat-generating equipment, they seek and apply alternatives to CFC gases for refrigeration equipment, they influence the use of materials in building structures which are non-pollutant to atmosphere in both their manufacture and application and they maximize the efficient use of our expensive energy sources. These are the essential areas of contribution by engineers to reduction of the greenhouse effect and the hole in the ozone layer of the atmosphere, of which we are all aware and afraid.

British consulting engineers have long experience overseas in adapting to rapid changes in political and economic climates. At home we are equally adaptable, so let me deal with our contribution within the United Kingdom.

Almost all renewal projects have a major engineering content. Environmental planning, land reclamation and protection, drainage, traffic access, transport systems, riverside and dockland development, and industrial and commercial developments are some of the areas where the skills of the engineer are of paramount importance. This is true whether the project is a new structure or the refurbishment of existing buildings and public services.

Throughout our country the demand for land space for development purposes is great but thankfully, at the same time, we defend those natural areas which make our country a green and pleasant land. It is essential, therefore, that the development potential of derelict or decayed urban land must first be recognized, environmental factors considered, and new attractive images designed and promoted. Only then will private investment be attracted and in these tasks the consulting engineer will provide invaluable help during the conceptual stages by providing the best technical and financial advice at the outset to ensure a sound and economic conceptual base for the use of such land.

The government is committed to the regeneration of inner cities but, while it is prepared to 'prime the pump', the major role in most areas of development is expected to be played by private sector interests. In past years it was usual for land reclamation, refurbishment and basic infrastructures to be largely financed by public funds and, in these activities, consulting engineers traditionally have been employed to provide technical support by the public authorities funding the projects. The same essential support will be required by private bodies undertaking inner city and urban regeneration projects, and consulting engineers are already working with many such private bodies. This area of work will increase during the decade which lies ahead as pressure for these areas of development inevitably increases.

Urban development and renewal cannot occur without good communications. The principal area of contribution by consulting engineers is inevitably in the transport systems of our country. Much needs to be done to improve the links between the major industrial and commercial centres of England, Scotland and Wales, and their direct linkage to the Channel Tunnel, the lifeline to Europe.

As for Europe, the opportunities offered are mind-boggling. Apart from those emanating from the development of the Common Market, another exceptional area opened up is Eastern Europe and the USSR. These countries require vast assistance from the UK, the western nations and the USA in the construction management skills which they lack. British engineers have for generations worked in the underdeveloped areas of the world. This will continue through the decades which lie ahead, but it is likely that our future contribution will differ from that of the past. Many of the so-called developing countries have, with our help, improved the education of their people and now have, in their own countries, engineering skills previously sought from elsewhere. The contribution of consulting engineers in future years is likely, therefore, to be more in an advisory and conceptual capacity than a 'hands-on' capability.

So you see, we live in a changing world and who better to meet the challenges of change than fresh young engineers? Engineers have by tradition been somewhat conservative and thereby slow to react to change, a fact noted by the late Ove Arup when he said:

> *You are right but that is what is wrong,*
>
> *You stand still but life moves on,*
>
> *To live is to change and not just to be,*
>
> *If you can't then your youth is gone.*

Many years ago when my son was streaming for his 'A' levels he asked me what career I thought he should follow. I told him it was for him to choose, not I, but in advising him on a path to follow I suggested that he should identify a career which he believed would bring him great pleasure and then examine it to see if it could support him. If it did, he should follow it, if it did not he should choose another career giving less pleasure and make the same examination. This iterative process should be continued until pleasure and support come together. I advised him never to make the primary choice that of financial support. I offer you the same advice and I hope that for many of you engineering will provide both pleasure and support. If you choose to be an engineer I hope that you will be able to look back as I do today and recognize that your choice was the right one for you.

Tom Smith

FREng, HonMConsE, F.I.MechE, FCIBSE, FInstE, FASHRAE

Educated and trained in Belfast and in 1950 entered building services engineering with consulting engineers J.R.W.Murland in Belfast. In 1955 he transferred to Dublin joining Varming & Mulcahy, the Irish practice of Steensen & Varming International, Copenhagen, and in 1957 came to London to open and direct a new practice, Steensen, Varming, Mulcahy & Partners.

This post lasted for 33 years during which time he was involved in many exciting projects around the world, including The British Library.

He became Chairman of the Association of Consulting Engineers in 1989, and in 1991 President of the Chartered Institution of Building Services Engineers where his primary object was to encourage awarenesss of the visual and aesthetic values in good design. As Fellow of the Royal Academy of Engineering he is greatly involved with the education and training of engineers, in particular in the appointment of Royal Academy Visiting Professors in Principles of Engineering Design.

A happy life

Jack Chapman

A happy life implies good health, the means to provide for a happy family, and interesting, satisfying and useful work, with congenial colleagues whose prime interest is also in their profession, rather than in making money. For me, a career in engineering has been consistent with those criteria.

I still remember when I decided to become an engineer. I suppose I was about 14; my father took me to see a heavy lift. Not heavy by present standards, indeed minuscule, but it was the girder that supported the gallery of a cinema. Perhaps a hundred people would be sitting above and below the gallery, and my father was responsible for the design. That seemed like an important and satisfying responsibility.

My parents would have preferred that I choose a more lucrative occupation, and one less prone to the economic cycle, such as medicine or the law – the latter on the grounds that I was prone to argument. So I had to study Latin. That may seem to be a *non sequitur*, but it was not so at the time. When it became apparent that my ambition was well-founded, they gave me total support and encouragement.

I enjoyed looking through my father's portfolio of drawings which he had done as an important part of his studies; they were drawn in Indian ink on cartridge paper, and the portfolio was two inches thick. They were more interesting than computer printout. He was taught by Professor G.F. Charnock, who was the author of four remarkable volumes on graphic statics.

Graphical construction still formed a small part of my studies, but now numerical methods have taken over, with the help of computers. I am constantly amazed at the power, and rate of increase of the power, of computers. The designers and programmers must be very clever. Yet they didn't think of the millennium problem. Or was it planned obsolescence?

Like all powerful devices, organizations, or individuals, the benefits of computers are accompanied by dangers. Previously, analysis required some visualization of the behaviour of the structure. Now it is possible to analyse without visualization, and the danger is that a wrong result will be believed. Computer output is never better than the input. Visualization, behavioural grasp, is just as necessary now as it ever was.

I completed my undergraduate studies, after some disruption, in 1942, when I was nineteen, and joined the Royal Engineers. Those four years were I think the best part of my education. I experienced levels of responsibility when young which would not have been possible in normal circumstances. I hope that we shall not find ourselves at war again, but I would urge any engineer (or architect) to seek site experience on graduation, as far from home as possible.

I then returned to Imperial College to undertake research, on a rather boring topic, as it happened. However, other more interesting projects came along, initially in connection with ship structures. At that time we were still a great ship-building nation. Ships are designed by naval architects; passenger ships have much in common with hotels, so the naval architect must embrace many of the considerations faced by architects. Oil tankers, gas carriers, bulk carriers, container ships, passenger ships, car ferries, all have special problems associated with the nature of their cargoes. The naval architect must cope with these problems, and with minimizing resistance to propulsion, with seakeeping, accident survival, as well as with structural strength. The structural considerations in ship design have much in common with those in design of steel bridges, offshore structures and other marine structures. They all require the services of structural engineers.

I particularly enjoyed investigations which related to immediate practical problems. I was of course only involved on the periphery, I did not have the satisfaction of total responsibility for major projects. But I gained a superficial knowledge of a variety of structures, and some appreciation of the decisions which those in industry had to make, and the problems which they had to face. I was approached by the chief naval architect of the Ocean Steamship Company, Marshall Meek; they had commissioned a new generation of six container ships. There was no prior experience of such large ships (they were 700 ft long) with such large deck openings, and they had suffered severe fatigue cracking soon after entering service. Not long after we had solved that problem, he telephoned me again. I still remember the conversation, perhaps because it was brief and to the point: 'We are thinking of building five new container ships; they will be 900 ft long. Do you think it will be alright?' 'Could you move the engine room further forward?' 'Yes, I think so.' 'Then I think it should be alright.' The ships were 105 ft 9 in wide (the maximum width permitted by the Panama Canal), and the deck strips between the closely-spaced hatch openings and the ship sides had to be minimized, in order to maximize the number of containers, which had standard dimensions. The main structural problem was that the large deck openings would reduce the torsional strength of the hull.

The project involved extensive design studies by the owners and builders, and model and fatigue investigations were undertaken, but the ships came into service with remarkable speed. The width of the deck on each side of the hatch openings was only 9 feet. As a precaution, the owners agreed to fix strain gauges at points of maximum stress, so that if permitted limits were exceeded the master could reduce speed or change course to reduce the stresses. After some time, the precaution was found to be unnecessary. The owner's responsible attitude is unfortunately not shared by all owners; the losses of bulk carriers are an international disgrace. The naval architect carries enormous responsibility. Marshall Meek, who in his career with the Ocean Steamship Company was responsible for the design of 99 ships, became president of the Royal Institution of Naval Architects, and is now master of the Royal Designers for Industry, a distinguished company drawn from all sides of industry, whose number must not exceed 100.

The number of sites worldwide for large arch dams is relatively small, but as with other large dams they can have a major impact on the national economy, ecology, population, and perhaps on the climate. Where rivers traverse more than one country, they can also have major political consequences. A failure can cause devastation to life and property. Arch dams are normally built in deep, steep-sided, narrow valleys. The water pressure is resisted mainly by lateral thrust against the sides of the valley. The Boulder Dam on the Colorado River is perhaps the best-known example. The Verwoerd arch dam on the Orange River in South Africa, was unusual in that the length/height ratio was greater than for any existing arch dam, so that the arch displacements could have an important effect on the vertical stresses at the base of the dam. The crest length was about 2,700 ft. The volume capacity of the reservoir is 5 million acre feet, that is, the equivalent of the area of England one foot deep. The designer was Dr Paul Back, of Sir Alexander Gibb & Partners, and he required a scale model of the dam with its abutments to be built and tested. This was a major undertaking; in a 1:100 scale model, the gravity stresses caused by water pressure and by the weight of the structure are only 1/100 of the full scale stresses, so these effects must be represented by radial forces applied to the upstream surface and vertical forces applied within the body of the model dam. This experience provided some understanding and appreciation of the problems faced by the designers and builders of these enormous structures.

Ronan Point was a multi-storey block of flats which consisted of precast concrete floor and wall units which were lifted into position and then connected together. The system had been developed and used successfully in Denmark. Early one morning a gas explosion occurred on the 18th floor, and detached the external walls from the floor at a corner of the building. Because the wall units were connected to the wall units above and below, but were inadequately connected to the floor units, the walls peeled away from the floors, which were not designed to act as cantilevers, and folded down on the diagonals. Tests showed that the wall to floor connections were sufficient to withstand wind suction, but not to withstand an explosion. The building regulations and the structural design code made no reference to gas explosions. The wall to floor connection detail had been published in the proceedings of a conference on system building at the Institution of Structural Engineers, but had excited no comment. Gas explosions in the UK are not uncommon; if the question had been asked of anyone 'should the effect of a gas explosion be taken into account in the design of buildings supplied with gas?' the answer would surely have been 'yes, of course', but the question was not asked. The engineer's ability to ask the right questions is just as important as

167

knowing the answers. The engineer's greatest problem is to think of what he or she has not yet thought of. The intention of risk analysis is to assist that endeavour. A particular difficulty is to allow for the effect of human failure in design or in operation of the facility. What was unforeseen can seem obvious with hindsight.

Another, greater, disaster occurred in the early days of exploiting the North Sea oil and gas field. *Sea Gem* was a jack-up drilling barge. The tubular legs passed through openings near the sides of the barge, which was raised by hydraulic jacks and held at the required elevation above the water by pneumatic grippers. The barge was raised or lowered by alternately operating the jacks and the grippers. The barge was suspended from the gripping collars by vertical tie bars. During lowering, the barge momentarily jammed on the legs, and then dropped a short distance until the fall was arrested by the tie bars. To visualize the effect, one has to imagine a hammer weighing several thousand tons. If the tie bars had been sufficiently ductile and tough they could have stretched sufficiently to absorb the energy of impact. In fact the bars were brittle and were roughly cut to shape, so there were large stress concentrations, and the tie bars on one side of the barge fractured. The barge fell sideways and capsized, with the loss of more than 40 lives. The divers' inspection found that the legs had broken in numerous pieces which were scattered over the seabed. It was apparent that the legs were also brittle. It was maintained at the inquiry that the brittle legs had caused the collapse because, when the tie bars on one side fell, jamming between the legs and openings would have occurred, and as the fall had not been arrested the legs must have fractured at that moment. This argument was correct on the assumption that the legs remained vertical. It was possible to show however that if the legs swayed sideways the barge could fall through a much greater distance before jamming would arrest the fall; the distance was in fact so great that the bending moment on the legs was large enough to ensure that failure would be inevitable regardless of the ductility of the legs. This behaviour was confirmed by means of a scale model, in which the legs were ductile, and the barge consisted of a solid steel block, in order to produce the correct stresses in the legs. The model was used in court to demonstrate the mode of failure. The barge duly capsized on the table and was precipitated towards the feet of leaping counsel, who were doubtless more learned as a consequence. The analysis and demonstration were important for two reasons: firstly, to enable due precautions to be taken in future designs, and secondly for contractual reasons.

Nuclear power was another new development during these years, and we were able to contribute to solving some of the structural problems which arose. Because the consequences of failure were so severe, whilst the system and structures were novel, every effort was made to foresee possible eventualities.

Another duty which falls upon engineers is the drafting or updating of design codes. A significant part of my time was spent on this activity, with the associated experimental and theoretical research, over a period of about 20 years.

My activities during those years were varied and absorbing, but I began to feel that I should experience a different environment, nearer the sharp end of construction. The first opportunity came when the British Steel Corporation set up the Constructional Steel Research and Development Organization, and I became its first director. After two years I was beginning to doubt whether it was possible to overcome the organizational constraints, when I received a letter from Sir Godfrey Mitchell, who had bought Wimpey, a small public works contractor in Hammersmith, with the help of his army gratuity in 1919, and had seen it grow to be the largest construction company in Europe. I was offered the job of group technical director; I said I thought that the appointment should be at main board level, and that was agreed.

At that time Wimpey had 21 operating departments in the UK, and Wimpey International was heavily engaged in the development of the Gulf emirates. Wimpey Canada had a wide range of activities, Wimpey Mechanical Electrical and Chemical designed and installed process plants, Wimpey Marine operated North Sea supply and towing vessels. Wimpey Asphalt engaged in surfacing roads and airfields; asphalt is a complex material, its behaviour being time and temperature dependent. There were large architectural and structural design departments, surveyors, and planning departments. Wimpey Laboratories employed 1000 people, provided central services for soil mechanics, structures and materials and engaged in land and marine geotechnical site investigations, and ground engineering.

In addition Wimpey had recently formed a joint company with the US company, Brown & Root, to design and construct oil and gas production platforms for the North Sea. A greenfield site had been purchased a few miles north of Inverness. Fabrication buildings had been constructed and equipped to shape and weld the connections between the large tubular members which are used for the supporting structure (jackets) of production platforms. A welding school had been established to train those without prior experience or to qualify those with experience. The largest dry dock in the world, with a 400 ft long floatable dock gate, had been constructed. The bottom of the dock was far below the water table, so the dock had to be continuously dewatered by pumping from a large number of tube wells sunk around the dock. The first two structures were to be placed in 400 ft of water, much deeper than any prior structure. The structures were built in the dry dock on a flotation structure which consisted of cross connected tubes each 30 ft in diameter. They contained equipment which would systematically flood the flotation structure on reaching site, where the jacket was to be launched under radio control by 'crash dive'. The components of the jacket had to be located in space with great accuracy, so that the small gaps between components which were to be filled with weld metal were within the required tolerance. This was achieved by means of precise survey stations which were located on surrounding hilltops. The project was not helped by the local authority which, having given planning permission for the construction yard, then refused permission for a labour camp, so for some time the workforce had to be accommodated in disused passenger ships, moored alongside, until permission for the camp was finally granted. When the structure was complete, the dock was flooded, the dock gate was floated aside, and the structure was towed to site, where it was successfully launched. The steel piles which fixed the jacket to the seabed were then driven. The whole operation, including construction of the yard, had taken about two and a half years. This was an astonishing achievement by the owners, BP, and by the US-British team of designers and constructors. I greatly admire those who were responsible, especially the site team, for their engineering and management skills, who were undaunted in the face of enormous difficulties. Since then there have been many other success stories in the development of North Sea oil and gas.

It is important, when entering a new industry, perhaps with the help of others, to ensure that the underlying technology is developed as a national resource, with a view to wider application. Many developing countries have been quick to realize the seemingly obvious, to our detriment. Shall we ever learn? Perhaps, when we eventually play our proper part in Europe, the blinkers will be removed.

Steel offshore production jackets are normally supported by tubular steel piles which are driven into the seabed. The piles are driven through pile guides, and through sleeves located near the seabed, both being attached to the jacket. The connection between the piles and the sleeves is made by filling the annular space with cement grout. The connection is effective because the grout is confined between the pile and the sleeve. In a particular development the ground conditions were such that driven piles were not suitable, and it was proposed to connect the lower part of the pile to the ground by filling the annulus between the pile and the ground by cement grout. The connection between the pile and the grout was assisted by weld beads spiralling around the pile. The total length of the pile was 176 metres and the connected length was 60 m. The pile diameter was 2 m, and the design load was 6000 tonnes. The designers were required to show that the pile-to-ground connection would have a resistance safety factor of 4.5, that is, that the connection would withstand 27 000 tonnes. The ground properties were given, and the specified radial stiffness was many times less than that of a steel pile sleeve, though the connected length was much greater; there was no prior experience of such a connection. The question put to us was whether the safety factor would be not less than 4.5; an answer was required in not more than 40 working days.

It was apparent that the confinement provided by the ground was vital to the efficiency of the connection, so it would be necessary to explore the effect of possible departures from the specified stiffness. It was also apparent that the shear distribution along the 60 m connected length would not be uniform. Because of the limited time available, four parallel activities were instigated:

- Laboratory tests on weld bead shear connectors in which the radial stiffness of the ground was simulated by springs of variable stiffness, to determine the strength and stiffness of the connection.
- Linear and non-linear numerical analysis to find the shear stresses and displacements at the pile/grout and grout/ground interfaces, for various ground and connection stiffnesses.
- Formulation of a simple design model for the resistance of the weld bead connection as a function of the grout strength, bead height, and the ground stiffness.

- Proposals for alternative connections which could be used if the weld bead connection proved to be inadequate.

We were able to demonstrate that for the given ground properties the resistance safety factor exceeded 4.5. The report was delivered on time.

Although the principal engineering institutions are named according to the broad disciplines of civil, electrical and mechanical engineering, it is important to recognize that most construction projects require the services of two or more disciplines; the development of offshore oil and gas fields, power stations and industrial plants are prime examples. Civil engineering by its nature is concerned with construction and it follows that civil engineers are, or should be, best suited to managing major construction projects. To do this they need to have a sufficient appreciation and understanding of the functions of other disciplines. The functional design of buildings is carried out by architects, and it is they who communicate with the owners, and who usually coordinate the disciplines of structural, geotechnical, mechanical and electrical engineering. The discipline of building engineering is also recognized. It is important that civil engineers have a broad understanding of and respect for all the disciplines which contribute to construction.

Why is it that the British media are apparently so disinterested in success? Is it because they are populated by journalists who live in their own little world of entertainment, sensation and trivia, and believe that the public share their own preferences and limitations? If that is their belief it can be self-fulfilling, because if the public are so fed, they will tend to demand more of the same. If children are fed sugar, salt, and fat, they soon regard those ingredients as essential. Whether or not this diagnosis is correct, it is important that engineers and their professional bodies regard it as part of their duty to communicate not only with their clients and themselves, but also with the public. We must not be satisfied with a situation in which engineering is only mentioned when disaster strikes. We must learn to communicate with the public and the media, and we must also address justifiable public concerns where policies or projects threaten safety or the physical or visual environment.

To communicate accurately, concisely and sometimes persuasively, the engineer must attain a standard and quality in written and spoken language which is at least as high as that attained in any other profession.

In the coming years, strategic environmental and economic considerations should become more significant in the design process. For example, the use of home-grown timber, perhaps in reconstituted form, assists the national trade balance. If tree planting keeps pace with or exceeds the rate of use, more carbon dioxide is absorbed. In designing water storage and supply systems, especially in developing countries, consideration must be given to the effect on the aquifer, on rivers, especially where rivers cross national frontiers, and on demography. Irrigation can have unwanted side effects, especially if combined with the introduction of chemical fertilizers. Recycling of materials will become more important. Developments in structural engineering will spring mainly from new requirements (for

171

example, in reaching deep ocean resources), new materials, and new production methods.

How can young persons decide whether their talents are such that they might become successful engineers, and whether the profession can offer them fulfilment? Ability in English, applied mathematics and science are basic necessities. Analytical and inventive talents are both required, but they need not both exist in the same individual. It is possible to discover or develop interests or talents before embarking on professional studies. Opportunities which exist in or out of school (or university) to develop inventiveness or managerial ability should be taken. The engineering institutions will provide introductory literature. Visits to installations, factories, or construction projects can be arranged.

Engineers, technicians, and craftsmen require different talents, but all members of the construction team are equally necessary. Most engineers will be pleased to talk to young people; a careers teacher should have a register of engineers who are willing to help. A criterion for entering any profession is that interest in the profession should be greater than interest in making money.

Jack Chapman

PhD, FREng, FCGI, FICE, FIStrucE, FRINA

Jack Chapman graduated from Imperial College in 1942, and following four formative years in the Royal Engineers, returned to Imperial College where he engaged in research, teaching, consultancy, and code drafting. He was appointed the first director of the Constructional Steel Research and Development Organization, and then became technical director of Wimpey plc, where his admiration for those who work at the sharp end was reinforced. He then founded Chapman & Dowling, and a variety of interesting investigations and design reviews ensued.

Following Patrick Dowling's appointment as vice chancellor of the University of Surrey, Jack Chapman continues to practise. He is a visiting professor at Imperial College, where his current research interests are in plated and cold rolled steel structures.

The Morecambe Bay Gas Field

Cedric Brown

Summary

This paper describes the work associated with the development of the More-cambe Bay Gas Field, the largest single project undertaken by British Gas (BG) and the first venture into bringing an offshore field into production. Particular attention is given to the energy climate in 1979 which led to the need for advancing and accelerating the development. Reference is made to the technical work from design through construction, commissioning to commencement of production from the offshore and onshore plant on 8th January, 1985, highlighting the various innovative techniques incorporated in the project.

The management structure and controls, essential for bringing to successful fruition a project of this magnitude, output up to 1200 mscfd (millions cubic feet per day), cost approximately £1.4 billion are identified. I conclude by setting out the achievements and most importantly, the lessons learnt, which benefitted the second stage of the development.

Discovery of the Morecambe reservoir

In the 1970s a major oil company was awarded exploration licences in the More-cambe Bay area of the Irish Sea. After investigations and seismic studies, it concluded that there were no commercially recoverable quantities of gas or oil in the area. It thus withdrew from any further interest. BG reviewed the data, disagreed with those findings and subsequently took up the options. The exploration effort which led to the discovery of the field is well covered in the technical press together with the various feasibility studies undertaken to assess the most viable methods of development.

By July 1978, the initial feasibility studies for the Morecambe Field were substantially complete and BG announced that the field was a commercial proposition and would be developed.

This is an edited version of a paper presented to the AGM of the Institution of Gas Engineers at Harrogate in 1987. It is reproduced by courtesy of the Institution.

Background gas supply and demand in the late 1970s

To appreciate the tight programme established by BG for bringing the gas field into production, it is necessary to explain the supply/demand position that existed in 1979. At that time there was the latest in a series of Middle East oil 'crises' which resulted in a substantial escalation in the world price of oil and reductions in output. The subsequent 'flight from oil', as it was dubbed by the Chairman of BG, led to a rapid build up in demand for gas in the United Kingdom.

As a consequence and the need for British Gas to retain control of the supply/demand position, a number of strategic decisions were taken, as follows, to:-

(i)   introduce marketing constraints to restrain further increases in demand
(ii) provide additional supply facilities, particularly aimed at increasing the amount of gas available to meet peak demand
(iii) reinforce the capacity of the national transmission system.

A programme of capital works was instituted to meet items (ii) and (iii) above including the development of the Morecambe Bay Gas Field as a peak supply facility, buy out and conversion of the Rough Field (southern North Sea) into a storage field, provision of further LNG storage, establishment of further salt cavities at Hornsea, construction of a fourth feeder from St Fergus into the north and Midlands of England and provision of additional compression capacity on the pipeline system.

The estimated cost of these works at 1979 prices was of the order of £4,000 million. Without the additional facilities outlined above, it was projected that there would be a deficit on supplies of gas.

While the general objective was to have all of these projects completed at the earliest opportunity, and thus be able to remove partially or in full the restraints on sales of gas, programmes were established for each project. A target of having some production available for the winter of 1984/85 was set for the Morecambe Bay development.

Characteristics of the field

The Morecambe Gas Field is located (*see* Figure 1) approximately 25 miles west of the Lancashire coast, with the reservoir an average of some 3,500 to 4,000 feet below sea bed level. Although the water depth in this area is only about 100 feet, there is a tidal range of the order of 30 feet which produces strong currents.

The reservoir, total reserves of about 6.68 trillion cubic feet with 5.26 of these being recoverable, is divided into two separate parts (i.e. southern and northern lobes). Each lobe is to be developed as a separate entity. The southern lobe comprises about 75% of the total reserves.

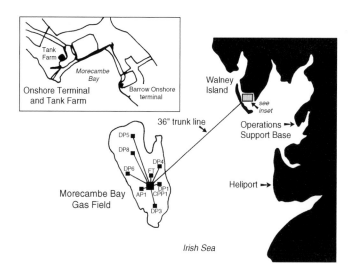

FIGURE 1 *Layout of facilities.*

The main constituents of the gas are:

Methane	84.0%
Ethane	4.3%
Propane	1.0%
Nitrogen	7.7%
Carbon dioxide	1.6%

The field also contains quantities of commercially valuable condensate, about 6 to 10 barrels per million cubic feet of gas.

Objectives of development

The objective set by BG was to bring about economical production of gas and condensate from the field for the 1984/85 Winter, earlier if possible, with the gas being to a specification suitable for feeding into the transmission system.

Development of the field would be split into two phases, phase I being the southern and phase II the northern lobe. Work on the southern lobe would be further subdivided into two stages with stage I constituting the southern section of the southern and stage II the remainder of southern lobe.

Stage I would comprise, as well as three drilling/production platforms, the central process platform designed to handle all of the gas from stages I and II, an accommodation platform, two jack-up slant drilling rigs, interconnecting pipelines, trunk line to the terminal, onshore terminal and condensate storage and pipeline connection to the national transmission system.

Production was scheduled to build up to 745 mscfd for the winter of 1986/87 from stage I and to 1200 mscfd by 1989/90 from stages I and II. Obviously actual production achievements would be heavily dependent on well productivities.

Initial planning and feasibility studies

Since it had been decided that the field would be used to provide peak load gas and hence operate at a low load factor (approximately 30%) then depletion would be over a much longer time span than is normal for gas and oil reservoirs, usually of the order of 25 years. As a result, a 40-year operating life was judged to be appropriate for the offshore structures and plant.

Because of the high cost of installing offshore facilities, it was decided to provide additional capacity on the central process platform over and above the 1200 mscfd (initially for standby but also to allow for further increases in supply) but onshore to ensure that space was made available at the terminal.

The chemical characteristics of the gas, wobbe number and calorific value meant that blending of the gas would be necessary to ensure compatibility with other sources of supply. Consequently, facilities were required for suitable mixing of Morecambe gas with gas from St Fergus at Lupton.

Offshore facilities

The comparatively shallow depth of the reservoir and, most importantly, this being the first offshore development by BG, underlined the importance of thorough study and evaluation. Two independent feasibility studies were therefore commissioned on the basis that the companies selected were expected to have different approaches to the subject. Options for development were left open, except that the necessity to spread the production wells over the whole of the field to ensure effective drainage was emphasised.

Alternative plans were presented at the conclusion of the studies in June, 1979, covering the various types and numbers of structures, method of drilling, and manner and stages of treatment of the gas for shipment to shore. After analysis of the plans, it was proposed that the ultimate development should be:
  i) Structures would be of steel construction. These are well proven in
      practice and in the water depths encountered at Morecambe, and no
      special technology would be required. Studies indicated that steel
      structures were cheaper and could generally be built faster than equiva-
      lent concrete platforms. In addition, the available soils data indicated
      that the seabed was not suitable for concrete gravity-type structures.
  (ii) Slant drilling techniques were proposed, in order to achieve maximum
      coverage of the field, while keeping the number of wellhead platforms to
      a minimum. Indications were that normal directional drilling techniques,
      commencing from the vertical would require at least thirteen wellhead
      platforms to provide effective overall field drainage of the southern lobe,

177

while slant drilling reduced this to seven (since reduced further in number). The studies indicated that the cost of drilling slant wells would be no greater than for conventional directional wells.

A study was undertaken of experiences in Trinidad, Peru and offshore California, and a report prepared on its suitability for Morecambe . Consequently, slant drilling was recommended as the optimum drilling method.

To minimise the cost of each drilling platform, it was decided to utilise jack-up vessels as drilling support tenders. These units would be used to transport the slant drilling rigs around the field and would provide power generation, mud supplies, accommodation for the drilling crew, storage etc. The drilling rig would be skidded from the jack-up on to the fixed platform for the drilling operation. Umbilical connections would provide the support facilities. This technique, like slant drilling, was something of a rare occurrence.

Site surveys

To establish sites for the onshore terminal, the offshore pipeline route, and the onshore support bases required during construction, and for long-term operations, detailed surveys were carried out to identify suitable locations.

Onshore terminal
The search for a suitable coastal site for the onshore terminal to process gas received from the Morecambe Bay gas field extended from St Bees Head in Cumbria to Anglesey in Wales. Safety and environmental factors were of prime importance with a requirement of about 200 acres (to provide for possible expansion in the future) and containing the following characteristics:

(i) Close proximity to the coastline and without public roads or railways between it and the sea.
(ii) A level area for development without major civil engineering work.
(iii) Reasonable access for roads, public utilities and gas pipelines.
(iv) At least a half mile from the nearest dwelling or industrial plant.
(v) Close proximity to a practical landfall for the offshore pipelines.
(vi) As free as possible from environmental and ecological restraints both onshore and offshore.
(vii) Minimum use of high grade agricultural. land.
(viii) Close proximity to the existing pipeline network and to large centres of gas demand.

Consultations with local government were put in hand, followed by discussions with specialised bodies such as the Ministry of Defence, Water Authorities, The Countryside Commission and the Nature Conservation Council. Consultations were also held with non-statutory bodies such as the National Farmers Union and The Country Landowners Association.

After consideration of all of these factors, a site was finally selected at Rampside. Following lengthy negotiations, planning permission was obtained.

While these developments were taking place, consideration was being given to the handling of the condensate. In-ground tanks were available within the Barrow docks and preliminary surveys indicated their suitability. Negotiations for that base were commenced with the port authority, and the local council. Attention to this aspect was necessary due to the Barrow Docks being used for the transportation of nuclear fuels and also the proximity of the Vickers Shipyard.

Offshore pipeline
Allied with the application for planning permission for the terminal site was that for the three miles of inshore pipeline from the low water mark on the west of Walney Island across the island, the Walney Channel and thence to the terminal. As this was being laid under the Pipelines Act it also required planning consent from Barrow Council.

Onshore pipeline
The planning, routing and wayleaving aspects and requirements of onshore gas pipelines have been covered fully in the technical presss and the pipeline between the Barrow terminal and the national transmission system was planned to comply with all of these criteria. However, at the southern end of the Lake District the terrain traversed by the line was most arduous and apart from a short distance at each end, did not permit conventional pipelining construction techniques; rather it became a series of special crossings to deal with the varied conditions.

Onshore support bases for offshore construction and field operation
During the initial stages of planning, it became necessary to give attention to the requirements for onshore bases to provide the management centre, materials storage and despatch, and the helicopter site for work that would be carried out offshore during the construction phase, and ongoing operations.

Final responsibility for selection of a suitable location rested with the exploration subsidiary of BG, the eventual operators of the field. Many criteria for the operations establishment were also common to the requirements for the construction site. The notable exception was that 24-hour access to port facilities was stated as an essential aspect of the operational base. Important factors that needed to be considered were:
- a port with adequate quayside facilities that would remain in operation over the timescale of the development;
- adequate loading/storage facilities;
- the availability or space for the provision of warehouse accommodation adjacent to the quayside;
- good road/rail access to the port;
- suitable office accommodation within relatively close proximity of the port;
- steaming time and hence cost between the port and field;
- good, reliable seaward access to the port;

- the nature of dock labour and alternative employment, history of industrial relations.

Following detailed assessment of all possible sites, from Cumbria to North Wales it was decided that the construction base would be at Barrow-in-Furness and the operations base would be established at Heysham. Helicopter facilities would be provided from Blackpool.

Offshore statutory approvals

Following the finding and proving of the field, it was necessary to obtain formal approval from the Secretary of State for Energy to develop the discovered resources. An essential requirement was that a Certifying Authority (C.A.) should be appointed to act on behalf of the Department of Energy's in respect of the Oil and Gas Enterprise Act 1982 and the mineral workings (Offshore Installations) Act 1971.

Organisational arrangements

The natural sequence of activities, offshore and onshore is as follows:-

Offshore	Onshore
Statutory approvals	Site selections
Design	Statutory approvals
Fabrication	Design
Transportation	Construction
Installation	Commissioning
Hook-up	
Commissioning	

The scope of work entailed divided naturally into the following areas.

Offshore facilities	covering the five offshore platforms and infield pipelines.
Drilling	including provision of the two jack-up units and slant rigs, as well as the drilling activity.
Offshore pipeline	provision of the 36″ diameter pipeline from the field to the onshore terminal (excluding the first half mile from the field – placed under the offshore facilities).
Onshore terminal	including the condensate storage facility and inter-connecting 8″ diameter pipeline.
Onshore pipeline	provision of a 42″ diameter pipeline from the onshore terminal to the national transmission system including the mixing station at Lupton.

Almost all previous projects undertaken by BG had been carried out within the functional framework of the organisation. However, in this particular instance, it was decided to adopt a 'task force' approach. This meant the appointment of various discipline engineers, accountants, planners, contracts material and procurement officers. The non-technical personnel were located within a central project services group.

Also attached to the 'task force' was a dedicated audit team, who had a direct reporting line to the manager of audit and investigations, and a senior financial analyst, who combined work on both the Morecambe and Rough Projects.

The drilling group had a responsibility for achieving objectives to programme and budget but retained a line into the drilling section of Exploration Department for technical support.

A special advisory group was established to coordinate the technical resources and interface with offshore management. The organisation arrangements are shown in Figure 2. Having decided upon the structure of the organisation, close attention was paid to the make up of the respective project groups.

A massive programme of capital works was being implemented and this placed a heavy demand on the in-house resources. In addition, it had to be recognized that this development represented the first venture by BG (on its own) into offshore project work. The experience within BG of such work was limited, and it was decided therefore to apply existing in-house resources to those areas of work on which our experience was established, onshore terminals, pipelines offshore and onshore, drilling and support services.

For the offshore facilities contractors were appointed to undertake management. The nature and scale of the offshore activity meant that there would be a good deal of forward planning and work of a strategic type, as well as an immense task of day to day detailed work. For this reason, and also to ensure closer involvement and integration between the contractor and BG, it was decided to appoint

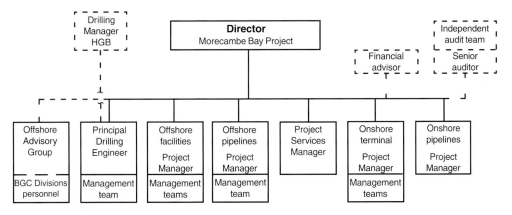

FIGURE 2 *Project organisation.*

two 'management type' contractors. The first would be for the provision of a project management team (PMT) containing and working under the control of the designated project manager for offshore facilities reporting to the director, Morecambe Bay. This small team would deal with strategic issues.

A second contract for management services (MSC) was to be awarded and this contractor would work under the supervision of the PMT. This group would provide engineers, planners, materials and contract staff, accountants, and clerical personnel to manage the day to day activities and was expected to reach in excess of 500 staff at peak.

One of the stated objectives was to ensure that over the duration of the project, BG personnel would be integrated within these offshore management teams; thus a number of in-house staff would acquire hands on experience of offshore work.

The scale of work at the onshore terminal meant that BG could not meet from within its own reserves all of the management team requirements. The project manager and other key project and site staff were direct employees, however, it was necessary to support this small group, by the appointment of a management services contractor.

A variety of consultancies and smaller support contracts provided specialist advice and assistance during key phases of the work. Amongst those were:

- consultants to advise and assist during the formation and establishment of the management structure;
- consultants to assist on the management of drilling activities;
- assistance on aspects of work associated with the offshore pipeline;
- management and design contract for the condensate storage facilities.

While this aspect of the project is primarily concerned with the management structure, it is appropriate to make reference to arrangements for carrying out design work needed both off and onshore. A prime consideration was with the numbers and experience of engineers/draughtsmen required. Major design contracts were, therefore, awarded on the basis of the following:

Offshore facilities	Conceptual design for offshore facilities.
	Contract for the detailed design of the jacket and topsides for the CPP and AP.
	Contract for the detailed design of the jacket and topsides of the 3 drilling platforms.
	Detailed design contract for overall control and communication system.
Drilling	Contract for conceptual design of slant drilling rigs
	Contract for detailed design and fabrication of the slant drilling rigs.

Onshore terminal	Detailed design and procurement contract for the entire terminal facilities.
	Contract for conceptual design of condensate storage facilities.
	Contract for management, detailed design and procurement of thecondensate storage facilities.
Offshore pipelines	Separate detailed design contracts were placed for a) offshore trunkline from field to shore
	b) in-field pipeline and risers.

Design arrangements for the onshore pipeline and mixing station were undertaken using in-house facilities.

Technical aspects of the development

30 million man-hours have been expended, as follows:

Offshore		*Onshore*	
Design	3.6	Design	0.5
Fabrication	9.1	Construction and commissioning	4.0
Hook-up and commissioning	7.9	Management	0.5
Management	4.6		
Total	25.2	Total	5.0

Many aspects of this development would require, and justify, a paper in their own right, for a full appreciation of the many technical components to be fully explained. However, an overview of some of the technical features is appropriate.

Offshore drilling

There are two noteworthy aspects of the drilling activity, namely, slant drilling and the use of jack-up tenders. The technique of slant drilling is referred to earlier and the advantages can be seen in Figure 3.

When drilling is required at any particular platform, then the jack-up would position itself alongside, and proceed to skid the rig off the jack-up onto the fixed platform. Final positioning on the platform would he achieved by moving the rig around the skid rails built onto the top of the deck structure. An interesting concept given rig weights of the order of 700 tonnes. A sketch of this arrangement is shown in Figure 4. The tolerances on the skid beams were very onerous at ±0.125 inches deflection.

Construction of offshore facilities

During fabrication of the jackets and top sides, contracts were placed with 13

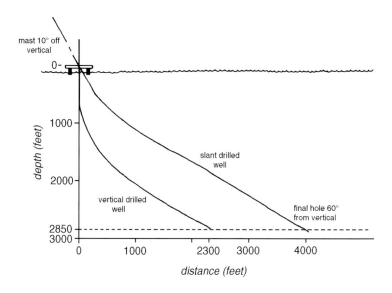

FIGURE 3 *Basic principle of traditional and slant drilling.*

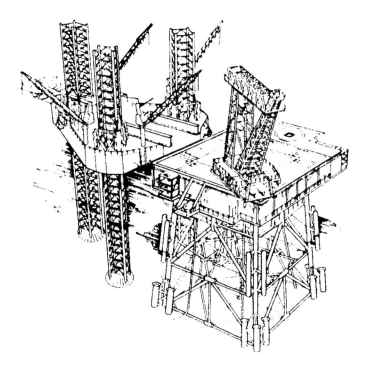

FIGURE 4 *Drilling platform (on the right) with jack-up support vessel alongside.*

yards. This, along with Rough, was the first offshore development where all fabrication work was undertaken in the UK.

It should be appreciated that while the jackets were each constructed as a single structure, the topsides were too large to be lifted as an integrated unit. It was necessary, therefore, to divide them into a number of modules. There were 34 separate items requiring movement from the fabrication yard to the field.

A resident team at each site of personnel represented BG and covered such disciplines as project management, engineering, planning, QA/QC, costs etc.

Transportation and installation

The 34 jackets and modules, together with steel piles, required transportation from the fabrication site to the offshore location. This was achieved by utilizing a fleet of barges. Since the majority of the construction was carried on the east coast of the UK, the sailing routes to Morecambe Bay were fairly lengthy and necessitated either a journey around the north of Scotland or through the English Channel. Weather conditions were critical in terms of achieving permission for the sailings. However, no units were lost, although there were a number of near misses such as during force 8/9 gales, when two barges broke away from their attendant tugs and had to be 'recaptured'.

The usual method of installing jackets is by either a straight lift with a heavy lift crane, or by launching from a barge offshore. In the former, a crane of sufficient capacity and able to operate in shallow water is necessary. For the latter, the jacket needs to be designed to withstand the stresses of launch. Jackets for the accommodation and central process platforms were installed by the lift and by the launch methods respectively. However, jackets for the drilling platform had to be installed using a somewhat novel technique. Originally, they were intended to be lifted into place, but as the weight of the topside equipment and slant-rigs increased during the design, the jacket weight reached a stage beyond the crane capacity. No other crane considered suitable for operation within the shallow and fast-current waters of Morecambe Bay was available.

The buoyancy effect achieved when a structure is immersed in water was applied here. In practice, this involved submerging the barge carrying the jacket until the effective net weight came within the lift capacity of the crane. This required submerging the barge almost 70 feet. A successful installation was achieved and the barge brought back to the surface and reused for two further installations. Extensive model tests (1/30 full scale) were carried out in Britain and Holland preceding the installation and during the design phase of work. A full scale trial of the submerged barge was undertaken in Rotterdam harbour.

Once the jackets had been located on the sea bed a series of steel piles were driven to a sound foundation. Following jacket installation, each module was lifted into position on the appropriate platform to form a fully integrated topside (see Figure 5).

FIGURE 5 *The central complex in its completed state* (courtesy British Gas Hydrocarbon Resources Ltd.).

Hook-up

On completion of installation, the programme of connecting all of the pipework, electrical, instrumentation and control systems could commence. This operation was basically a large logistics activity, since it was necessary to arrange, first of all, for the delivery, to an onshore location and thence to the field, of a wide range of equipment, materials, consumables etc. required to complete the installations. The hook-up activity involved an offshore labour force in excess of 1200 at peak, and at completion had absorbed about 8 million man-hours of effort.

Matters of particular technical interest

Control system
A sophisticated and advanced control system was developed for offshore, based on normally unmanned operation of the drilling platform. This computer-based system was initially designed for four levels of control, subsequently modified to three.

    Level 1 (the one to be discarded) provides for the control from the onshore base at Heysham.

    Level 2 provides overall field control from a purpose-built control room on the accommodation platform.

    Level 3 provides process control from the control room on the CPP.

    Level 4 gives local plant control on each platform.

Each level is capable of complete control of the level beneath.

Waste heat recovery
Power turbines located on CPP provide power supplies throughout stage I. The waste heat is used to heat the oil used on the glycol regeneration process.

Accommodation
Full accommodation, catering, housekeeping and leisure facilities are provided on the accommodation platform (AP1) for 178 persons. A full scale mock-up of a bedroom/lounge area was produced during design and reviewed for acceptability by the field operations staff. In addition, this platform contains the field control facility, thereby segregating it from potentially hazardous plant areas. Also incorporated is a full communications facility, together with hangar and fuelling facilities for a helicopter.

Special inhibiting occurrences offshore
Of interest are the following:
• incomplete design at commencement of fabrication;
• delays and changes arising from the above;
• modules incomplete upon installation in the field;
• discovery, when offshore, of defects in several hundred valves, necessitating an extensive testing, refurbishment and replacement programme;
• discovery of cracks in the jacking system of the accommodation vessel *AV1* being used as an offshore hotel for construction workers, leading to its removal from the field.

Each of these aspects needed careful attention by management in order to retain control of the programme and tests.

Offshore statistics

Some of the statistics are also important in understanding the scale of the activity:

	CPP	*AP*	*DP3*
Jacket weight (tonnes) including piles	11 754	4100	5352
Topsides weight (tonnes)	12 933	5394	5433
Total weight	24 687	9494	10 785

It is also estimated that the following quantities of materials have been expended offshore:

Electrical cable	1800 miles
Stud bolts	650 000
Valves	36 000
Steel	70 000 tonnes
Piping	200 miles

The scale of transportation of personnel to the field by helicopters is shown by the following:

Personnel movements by helicopter (weekly)	2500
Personnel movements by helicopter (from commencement to November 1986)	326 000
Helicopter flights (weekly)	370
Helicopter flights (from commencement to November 1986)	14 500

Offshore pipeline

Construction of the 36″ diameter offshore pipeline was divided into two separate contracts because of the different laying techniques required. The offshore section of pipeline west of Walney Island lent itself to traditional submarine pipeline construction, using a lay-barge. However, the channel section between Walney Island and the onshore terminal required a different approach as the use of a lay-barge was impractical due to the shallow water and a profusion of mud flats and tidal areas. This necessitated the use of a dredger for trenching with the pipe being winched, in sections from onshore.

Construction of the offshore section of the pipeline began in April 1982 using a flat bottom barge with centre stinger. However, because of shallow waters, this vessel could not operate close inshore and the first mile or so west of Walney Island had to be installed by pulling pipe from the lay-barge using island-based winches.

Onshore terminal and condensate storage

Following commencement of design in 1980, construction work on the onshore terminal (Figure 6) commenced in April 1981 and at peak in August 1984, over 1,500 personnel were employed on site. A total of 54 contracts were let, plus numerous small orders for construction services, vendor supervision and commissioning.

A much smaller but still important project was concerned with the refurbishment of tanks in Barrow Docks to provide condensate storage. Five of the ten tanks were leased, re-roofed, repaired internally where necessary, and all piping, electric and instrumentation re-designed and replaced. A jetty and tanker loading capacity was also required.

The following statistics of materials used on the terminal may give some indication of the scale of operations:-

Length of cable	370 miles
Length of pipework	25 miles
Weight of major machinery/plant	3896 tonnes (221 items)
Weight of plant steelwork	1748 tonnes

FIGURE 6 *The onshore terminal at Barrow* (courtesy British Gas Hydrocarbon Resources Ltd.).

Onshore pipeline and mixing station

Gas supplied from the Morecambe Bay Field is then piped to the Lupton mixing facility and blended with Northern Basin supplies before entering the national transmission system. This involved the construction of an onshore pipeline which passes through one of the most scenic areas in England and therefore presented a major environmental challenge.

Construction of the onshore pipeline commenced in April 1982. The route passed through a variety of landscapes such as agricultural land, pasture, Forestry Commission and National Park land. The ground conditions varied from rock, sand, gravel to silt, clay and peat. Five rivers, one canal and numerous streams, two railway lines, one motorway and 64 major and minor roads/lanes were crossed. A considerable length of the pipe trench was constructed through rock which necessitated controlled blasting using approximately 60 000 lbs of explosive, and removing 10 000 cubic yards of rock. The deep alluvial silts and clays in the Leven and Kent estuarine areas, extending for a length of some seven miles proved especially difficult.

Project services

All the necessary support services of finance, contracts, material procurement and storage, planning and administration were brought together in one coordinated team. This group provided a common service to those parts of the project managed by BG personnel and also carried out a monitoring role of the managing contractors.

In addition, the project services group acted as a staff arm to the Directorate and provided regular reports on progress against programme and budget.

Considerable attention was given to award of contracts, in particular, the major items of management and design. In each instance thorough and detailed presentations were called for and interviews held with senior management and proposed project personnel.

Controls and reporting

Up-to-date and regular information was essential in ensuring control of the project. Considerable use was made of computer-based facilities to gather, collate, compare and present information on programmes and costs.

This information was prepared in various levels of detail for the management concerned with the many facets of the total project, offshore and onshore.

Comprehensive reports were used as the basis for the regular coordination and review meetings which were held. The main meeting, held monthly, from commencement to completion, was the project-coordination meeting, attended by all project managers, project services group representatives and the managers responsible for onshore and offshore operations. This was used for presentation and discussion of reports on particular activities and difficulties, as well as reviewing programme and cost. Action lists were produced from the meeting and reviewed at the following meeting. Weekly change control meetings were held to review and reject/approve proposals for changes to the plan and technical details. Additionally, separate and independent risk analysis studies were instituted as well as a number of audit checks.

A comprehensive computer-based archival system was established early on in the project, to handle the large amount of documents on the offshore side of the project. Recording, location and retrieval of the many certificates, design proforma etc. was necessary to ensure achievement of certification and assist in the commissioning and handover. This involved in excess of 4 million documents.

Use was made of incentive schemes. One such innovation was with the offshore management services contractor. For a defined period of activity over a time of almost two years a pre-estimate of management man-hours was agreed. For each man-hour saved, and using an average cost per hour, the contractor gained 50% of the saving with the other half accruing to BG.

Of the contractors' bonus, one half was paid during the month of achievement with the remainder being banked for payment if the end milestone of first gas was achieved. A date of 31 December 1984 was agreed. In the event, first gas occurred on 8 January 1985.

Achievements

The project was launched in January 1980. By April 1981 the first ground was turned on the site of the onshore terminal. On 10 July 1983 the first offshore structure was placed, being the jacket for DP3.

The first well was 'spudded' on 26 August, 1984. After nearly 30 million man-hours of effort the Morecambe Bay gas field was brought into production on target, on 8 January 1985, in time to help meet the peak demands that followed.

With further increased production available through the 1985/86 winter, stage I of phase I was completed, except for drilling of the final wells by November 1986. Production in the winter of 1986/87 passed all targets and in excess of 950 mcfd was produced during the cold spell of January 1987.

The various techniques, including slant drilling, have performed well. It is worthy of note that the time per slant well was estimated originally at 53 days and to date has averaged at about 56/57 days per well completed.

Lessons learned

On projects of any size and type, but more particularly one of this magnitude, critical examination of performance will identify scope for improvement. Such a review has been undertaken of the Morecambe Bay project, and a number of significant lessons identified. The major ones are:

- use of in-house personnel to lead the project team. It was found that in general, management contractors were reluctant to direct others and hence assume responsibility for decisions.
- use of an integrated management structure.
- completion of design before commencement of construction.
- improved monitoring of performance in fabrication yards.
- completion of all possible work (including commissioning) onshore, prior to transportation offshore.

However, it must be realised that the particular project documented in this paper was set an extremely ambitious target and inevitably the areas of concern identified above were almost certain to arise.

One of the most significant lessonsgained is that the fundamentals of engineering and project management work offshore are no different from onshore activities. There are, of course, some areas which are particular to an offshore situation, namely the scale and expense of installation, the logistics in organising and moving personnel and materials during hook-up and costs being much higher offshore than for similar work onshore.

However, application of the techniques and disciplines developed from an extensive onshore construction programme has demonstrated that many myths about the offshore scene being more demanding, more difficult are just that, myths.

Any large project, onshore as well as offshore requires careful planning, sound controls and firm management.

Costs

Total costs can be approximately analysed as follows:-

Sector	% of Total Cost
Offshore facilities	67
Drilling	10
Offshore pipeline	3
Terminal (including condensate storage)	8
Onshore pipeline	2
Project services	9
Others	1

It is also worth recording, that approximately 86% of expenditure occurred in the United Kingdom.

Commissioning and operations

Throughout the duration of the project, particular recognition has been given to the interests of the departments and staff who would assume responsibility for subsequent operation.

Concluding remarks

Above all, however, this project is about teamwork. Successful implementation and operation of a multi-disciplinary task force was a critical factor in achieving production on target.

Acknowledgements

I would like to take this opportunity of acknowledging the quite exceptional efforts made by very many people in bringing this project to a successful completion. From within BG, its many contractors, managing, design and construction, supply and manufacture, the effort has been in the main excellent and well over 80% of this has been UK engineering effort.

Finally, may I thank the Chairman and Managing Director, Production and Supply, together with the Board for their support and permission for this paper to be presented.

Cedric Harold Brown, FREng

Cedric Brown's career started as a laboratory assistant with British Gas East Midlands in 1952. He trained as a pupil gas engineer and held a variety of posts in East Midlands before being appointed Chief Engineer and then Director of Engineering.

In 1978 he moved to London to take up the appointment of Assistant Director (Operations) at headquarters and a year later he became Director (Construction). Director (Morecambe Bay Project) was his next appointment where he was responsible for the development of the Morecambe Bay gasfield and also for pipelines, plant, construction, communications and instrumentation activities.

In 1987 Cedric became Chairman of British Gas West Midlands and in 1989 he returned to headquarters as Director of Exploration and Production and Member of the Executive. He was appointed to the Board as Managing Director of Exploration and Production later in 1989 and in December that year he became Managing Director, Regional Services while still retaining responsibility for Exploration and Production until June 1990. In April, 1991 he became Senior Managing Director and was appointed Chief Executive in 1992. He retired in 1996.

He was elected to The Royal Academy of Engineering in July 1990 and is a Fellow of the Institution of Civil Engineers, Fellow and Past-President of the Institution of Gas Engineers and a Liveryman of the Worshipful Company of Engineers.

Born in Portsmouth, Hampshire but with formative years spent in Yorkshire, he was educated at Rotherham, Sheffield and Derby. He is married to Joan and they have three daughters, a son and six grandchildren. His interests are family, sport, the countryside and reading.

A blinding glimpse of the obvious

Dudley Dennington

*A*t the age of 38 I was faced with a career dilemma. In a good job heading up the structural design department of Wimpey, I had the opportunity to apply for the post of assistant chief highway engineer for highway construction, in the new Greater London Council under Peter Stott, director of the new department of highways and transportation, with whom I had had no contact up until then. Incidentally, all titles of staff and departments in the GLC were spelt without capitals in all documents, a sensible arrangement that avoided aggrandisement.

I saw Peter and was inspired by his enthusiasm for the work ahead, but being uncertain of the life of a local government officer I decided not to move. About two days later I had the irresistible impulse to pick up the telephone, speak to Peter and ask if the position was still open. It was; what is more he was about to report to committee that afternoon that no suitable candidate had been found for the post and he was proposing to combine it with another already filled by a former London County Council man. I could not have left it a moment longer, and it confirmed my view that occasionally in life things are just meant to be.

It was a committee appointment and I was interviewed in a dark panelled room at County Hall sitting on a chair in the middle of a room with a dozen or so people at a table that curved around me; Peter was the only one there whom I knew. I had never been in charge of highways so I wondered how much my ignorance showed. After being asked to withdraw, I waited in another room for an uncertain time until a bell rang and a messenger fetched me back.

The chairman, now a well-known Labour peer and solicitor, started to tell me of the importance of the job and from his tone I was certain that it was a polite letdown; with surprise I was told that the committee were to recommend me to the all-powerful general purposes committee that afternoon. I learnt afterwards that you were only called back if you had the job. It was the same chairman in the afternoon; I had been warned to say a brief humble thank-you, and when confirmed in the post subject to a medical check, I went straight up to the doctor to be pronounced fit. Armed with a letter of appointment a few days later all I had to do was tell Wimpey. A good man was appointed in my stead.

Going to work at County Hall from Wimpey was like going to live on another planet, my tasks were so different; let alone the new aspect of highways, I had to attend committees and respond to politicians whilst I was leading engineers who were steeped in such activities. I was a new boy at a senior level and had to prove myself.

On my second day the finance committee met with city financiers in the minority party keen to ask incisive questions drafted by their accountants. Attending to answer for matters of expenditure in which I had played no part was going in at the deep end with a vengeance. My administrative officer suggested that my predecessor, now in a parallel job of equal status, could attend for me, as he had handled the matters concerned. I said no, it was my job and I would do it. It was the steepest of learning curves; I learnt the facts, reported them and survived, with prompting from my administrative officer sitting behind me.

Organization was the first task. Formation of the GLC had meant the extinction of the Middlesex County Council as an administrative body; their county surveyor, Stuart Andrew, was deputy to Peter Stott and I had to meld their highway team, still in their old offices, and the former London County Council (LCC) bridges section into one department for highway construction, or improvements as they were called; neither side were in a fully cooperative mood. My ignorance, however, was a benefit to me; I was not too sensitive and felt that if I could encourage them with the excitement of the workload ahead, and give them plenty to do, then the majority would come along and bring the rest with them. Fortunately I had one or two newcomers from the outside on my staff to help things go my way.

My first two steps were to get the Middlesex team, who were feeling sorry for themselves in Vauxhall Bridge Road, into County Hall alongside those from the LCC, and then to switch their leaders; they had to work together then. Each leader rolled up his sleeves to show his new lot how things should be done, and each team saw that honour had been done to their boss.

It was a holding position to make sure that the contracts in hand by the LCC and Middlesex County Council kept going smoothly under GLC adoption, while I had the opportunity to observe individual strengths and weaknesses and decide how to run my 30-strong division, looking after twice as many miles of major roads in London and the planned urban motorways and ringways.

I soon noticed a completely different approach to responsibility between the private and public sectors. In the first it seemed that you had to guard your authority, others were only too keen to take matters out of your hands if you faltered, or were not around at the right moment. If a new task arose in local government the first question the old guard asked themselves was whether it was someone else's job to handle; it was incorrect to usurp. The other big difference was the fact that instead of being given projects to design and construct, you had to decide what needed to be done in the first place. You planned the workload. It took patience and effort to sort things out, but by the time I had been upgraded to chief engineer status, taking on highway maintenance, I had a workable organization in place.

A word about Peter Stott, who had a significant effect on my career. He was my age and a big man in all respects. He had left an international consultancy in which, among other things, he designed bridges, to join the LCC. He also thought big. I had not met his kind like before, and in the seven years I worked with him there was never a reason for us to fall out. He guided with good sense and left you to get on with things; he saw ahead and around; and had immense drive instilling enthusiasm. A realist, one of his expressions when up against a problem – which I have since adopted – was the need to seek 'a blinding glimpse of the obvious'.

I had been fortunate to have had other distinctive mentors in my career. In my first job with a relatively small consultancy after leaving the army, the partner to whom I worked was Major Zinovieff, of Russian émigré stock, who had served in the Royal Marines and married into the British aristocracy. He too was large, expansive and friendly, being fully supportive. In Wimpey it was Dr MacGregor, director and former university lecturer; he had the gift of giving you his full attention when you discussed something with him.

The army had given me a good grounding in the need to look after your troops. These three mentors added to this, not by instruction but by example, namely: give generous time to people; see your staff and go to them; and always be cheerful when with them – there is no incentive to get on if it means you end up miserable and ill-tempered like your boss.

I found that there was a very interesting person at a relatively low level on my staff, he was a brigadier and a sapper, one of Major-General Wingate's Chindits who had set up base behind enemy lines in Burma. Considering that this distinguished soldier with DSO and bar was under-employed, I took the training of graduate engineers out of the hands of the administrative staff and gave it to him – not without the struggle as such storming of old bastions involved. One of his sayings was that there were three factors that contributed towards someone staying in a job: salary, job satisfaction and future prospects. If a person was satisfied with two of them they would stay, if satisfied only with one then they would most likely leave. I believe this.

We gave the graduates a good training, exchanging them with other firms, as well as with the Port of New York Authority. In particular we instilled in them the four basic parts of an engineering report: specified requirements; relevant factors; options; and justified choice.

Dealing with the well-established administrative staff was quite a task. They certainly knew their work, especially preparation for committees, the narrow door through which all decisions had to pass, and how to deal with the members who made important and expensive decisions on your advice. I had an excellent administrative officer, totally loyal to me and patient with my original ignorance of the workings of local government; he had joined the LCC more than thirty years earlier. He made sure that I was never liable to surcharge, and when, as happened, some calamitous situation arose he was Job's comforter saying that 'things were likely to get a lot worse before they got any better'; how true.

There was a large programme of work ahead of us, as well as much in progress from the LCC days. I had to settle claims on the driving of the second Blackwall Tunnel and prepare the documents for, and let, the last contract for its completion; basically its fitting out including the carriageway and wall lining. We generally aimed to do half of our work in-house and half through consultants under my control; this provided just the right edge of competition to get the best out of the GLC staff.

These were the days well before fee competition and we had no list of approved consultants. If one asked to go on our list I explained that we did not keep one: when we considered we required a consultant's services we approached the one we considered to be most appropriate – a reply which covered everything. As all such appointments required committee approval, as did the tender lists for contracts, on my advice, I tended to find myself quite popular with consultants and contractors, but always played a straight bat.

Whether it was the sheer interest and the scale of the department's work, or whether those outside the profession did not appreciate the ability of engineers to manage, there were one or two attempts to handle matters, other than engineering, for us. Having wrested the training of graduates from the administration branch, I had also to deal with the establishment branch and certain committees, who questioned whether contracts would be better supervised by administrators. I resisted all attempts politely but firmly, explaining the good sense of traditional arrangements, whilst making sure that we demonstrated this.

Whilst ensuring that the engineers were handling their own work I had to organize them beyond the initial sorting out in order to continue, without let-up, the jobs in hand from the Middlesex County Council and LCC and tackle the far larger future workload.

I am of the view that designer of a scheme is not the best person to manage its programme and progress; they count the months in hand like Tasmanian Aboriginals once did: 'one, two, plenty'. I therefore set up a breed of project engineers to oversee all aspects of progress, as public works involved consultation with numerous parties inside and outside the Council on top of the design. I ran into some unexpected opposition from those chosen; they did not appreciate losing the cachet of a team under their direct control. This was overcome by the ingenuous suggestion that perhaps I should give project control to the administration branch; opposition melted.

Despite views from outside, local government is not impractical. Because the Victoria Embankment supported a metropolitan road – part of the enlarged 800-mile network for which we were responsible – I had the job of looking after seven miles of all such embankments such as the South Bank. Also, because it had a pipe subway in it I looked after nine miles of these as well with a team of inspectors; add thirteen bridges over the Thames, and about 130 others, three road tunnels under it and two foot ones, there was enough to maintain.

Peter Stott was a good communicator; he held lunch-time meetings of his 'cabinet' at which we all aired opinions concerning the present and future handling of the department's work. The researchers were at one end of the spectrum and I was at the other, with the planners in the middle – from the ultraviolet to the infrared.

He left me to handle the finance committee work which was mostly mine; I was spending most of the money and had some awkward cases to put to them on occasion when costs on contracts went up. The solution is always to attend the committee even when matters are trivial. People are human, and if they get to know you during fair weather they treat you more kindly when it gets rough; and it certainly did sometimes. You had to be explicit and patient, remembering that communication with lay persons is a vital part of an engineer's job.

One thing is important when you secure a vote to spend large sums of public money: make sure you get on with things and do so. Otherwise you will be rightly criticized, and furthermore lose credibility during the next round of estimates when seeking funds. Also, arrange opportunities for those who agree to the expenditure to see what is happening on the ground occasionally.

I sometimes used a bit of stage management to achieve my ends; it was not necessarily deliberate. Road maintenance always required a large and easily identifiable sum to be included in the revenue estimates, from which it was very tempting to make cuts for the Council's budget, and I had to negotiate hard with the vice-chairman of the finance committee to keep it, providing good reasons for its need. At one meeting I took along a whole bundle of photographs of potholes and the like in metropolitan roads; I only showed him the top one as there was a lot to discuss, but later he said that he always remembered all those photographs I showed him!

There was a particularly obvious deformation in the road outside County Hall which I treasured, pleading general lack of funds when members asked me what I was doing about it. I was not pleased when my staff arranged its repair without telling me.

The finance committee and the treasurer's department who served them were powerful; all reports to any other committee involving expenditure went to them for concurrence; they were of a mind that other disciplines were not expert when it came to money matters. Grant, financial support from government funds for road improvements, was a case in question. Basically although the GLC had the powers to carry out highway works paying all themselves with money from the boroughs, it was a *sine qua non* that they would not do so if 25% was available from national funds. To what parts of a road scheme that this grant extended was a matter of tight negotiation with the Ministry of Transport.

If we went to the finance committee with the result of what had been extracted from the Ministry, and they instructed us to see if we could do better, the result of going back having squeezed some more out of the Ministry was not praise but censure; it was clear to the committee that we did not try hard enough the first time; that we were not up to the task of financial negotiation; and that it would be better if the treasurer took over the work. No wonder that when the Ministry officials dug their toes in we asked if they really meant it, and if so never to change their mind.

From the time of my appointment as chief engineer I was named as the engineer in the conditions of contract for the highway works we carried out ourselves, about £200 million a year at present-day prices. It was an interesting situation whereby I could make decisions binding upon my employer, the Council. No conflict arose.

Soon after my upgrading there were two major changes. London's traffic needed a supremo and Peter Stott was designated traffic commissioner, a post he held alongside that of director of highways and transportation. This was followed by an amalgamation with the planning department and the formation of the department of planning and transportation of which Peter was a controller, with a separate chief officer post of traffic commissioner and director of development filled by Alex Morrison, to whom I answered.

Alex was simply a born manager, and a delight to have as a boss, a very sound person who gave out confidence; he taught me a lot about handling matters and people. He had run munitions factories at an early age during the war, and went on later to be the first chief executive of Thames Water, and then chairman of Anglian Water. He left my scheme of organization alone and melded the planning architects into the enlarged department alongside the traffic and road planners.

The reason for the amalgamation was simple. Two separate committees dealt with major road schemes, highways and planning, with officers from our department, architects and planners confusing them with conflicting environmental opinions. Far better to thrash it all out before reporting to a joint committee, and with people like Peter and Alex it worked.

While all the organization side was falling into place, committee work handled, and funds secured and managed, there was still the engineering to do. Urban highways are not easy, the statutory side alone is very involved, but once the land has been secured and orders made, and with as many pipes and mains – hopefully in their advised positions – diverted in advance, it is just a matter of doing the work to a design which can be realized with the minimum disruption to the traffic which cannot be stopped. Greenfield sites rarely occurred.

Tibbets Corner underpass on Wimbledon Common was the first scheme to be constructed by the GLC, albeit it was authorized under Parliamentary powers sought and obtained by the former LCC. Many followed apace, those promoted after the GLC was formed being carried out under their new highway powers.

These were days of the motorway box and Blackwall Tunnel southern approach was the beginning of the plan, and London's first section of urban motorway. On one day we arranged with ceremony that the second Blackwall Tunnel was opened, the old one closed for refurbishment; and the southern approach started. Not easy to arrange, but exactly what our political masters liked.

The environmental problems and protests which occurred when the elevated Westway Road was opened about three years later were absent when the Blackwall Tunnel southern approach was built. The resident engineer kept in touch with the local people throughout and received a presentation from them when the road was opened. During the construction of the elevated Westway I had the honour to show Prince Charles around the works together with the consultant's resident engineer and the agent. At one stage when we were under part of one of the massive precast viaduct units, there were indications that the demonstration of lifting it off the ground was about to happen ahead of schedule; shades of the Tower of London loomed, and I moved the party on very quickly.

In 1970 Alex Morrison was needed for other things in the Council and the traffic commissioner post and director of development became vacant. I was short-listed for the post along with a senior architect as it was always seen as a manage-rial position. When the bell rang from the interview room, and the messenger's steps were heard coming down the corridor to summon the successful candidate, it was a relief to hear my name called. I was a chief officer.

Up to that point in my career I had always been in charge of people whose job I could do myself with some degree of success. From now on I was still the Coun-cil's engineer but I had architects and planners to direct, as well as a team working on computerized traffic signalling. We were the traffic authority for all 6000 miles of roads in London and environmental awareness was growing. The overall task was very different from the one for which I had trained as a civil engineer – and why not. The construction branch I had set up needed no alteration, only a new leader, and we were fortunate to have Leslie Deuce from the Ministry of Trans-port eventually in post – a splendid bridge man and civil engineer. The rest of the 800-strong team were divided on a geographical basis and a functional one.

A management consultant could have shown on paper that the geographical organization was unworkable, but he would have been wrong. There were five area teams, each a mixture of engineers and planning architects, dealing with all the planning of all schemes in their areas and their environmental issues; each had an engineer and a planning architect as its joint head. The five teams answered to three managers – an engineer, an architect and a planner – who in turn reported to me. Frequent semi-formal and ad hoc meetings ensured that all went smoothly.

Traffic management, particularly signalling, and making traffic orders, were functional activities. The first was innovative, the second routine, as was dealing with planning applications and administering the high building policy for Lon-don; this latter task involved making sure that buildings near St Paul's kept within the 'saucer' structure gauge for height so as not to overshadow it, and seeing what effect other buildings could have on London's skyline from near and afar.

The signalling was a major operation. London had 1500 traffic signal installa-tions, 30% of the total number in the whole country, which is not too surprising when you realize that the next thirteen of the largest cities in Britain would all fit into Greater London with room to spare. Following the success of computer link-ing of 90 installations in west London, it was decided to proceed with another 1000, mainly central.

The police control traffic, and there are two police forces in London, the Met-ropolitan looking after 640 square miles, and the City with one square mile. Set-ting up the arrangement whereby there were three management consoles for the linked signalling system housed by the Metropolitan Police but manned by them, the City Police and GLC staff was not a facile operation, neither was getting each party to pay an equal share.

As we were the order-making authority and the police were the enforcement authority, and the London Government Act required us to consult them on all draft orders, a forum for this was set up on a standing basis which had the wider interest of cooperating on all traffic issues. This was the Joint Traffic Executive for London of which I was a co-chairman with the then assistant commissioner for traffic at Scotland Yard, assisted by a few senior staff from both our organizations; we met monthly alternating between Scotland Yard and County Hall, each chairing the meeting according to location. It worked well and I could field new ideas like closing Oxford Street to all but buses and taxis well in advance so that it went ahead smoothly. Again, it was an interesting job for a civil engineer.

There were other diverse posts I had to look after such as chairing London's first road safety unit, and meetings with the Association of London Borough Engineers.

It was pleasant to find that a friend from Wimpey had also joined the GLC, Les Hatherly, who made a brilliant unassuming job of looking after road maintenance. An expert in asphaltic materials, his study of skid resistance led him to cooperate with Shell to produce Shellgrip, a highly skid-resistant surface dressing comprising calcined bauxite chippings secured by epoxy resin adhesive. It was a winner, reducing shunt accidents by 25% where applied, and it went worldwide. With 60% of all the country's personal injury accidents occurring in towns within 20 m of a pedestrian crossing or road junction, it proved excellent value when applied, as it is to this day; any spare money I had went into treating blackspots with the dressing according to Les's priority list.

The work was diverse and intense but at least I knew what I had to do. Apart from the engineering, internally it required leading at committee every fortnight, when the Council was in session. The process started on a Friday, meant reading all the reports over the weekend and then, after orchestrating which officer should speak about details at committee if necessary, attending on the following Monday after briefing the chairman; the next day involved the finance committee.

Externally, apart from dealing with the police, there were occasional ministerial meetings when the leader of the Council met the Minister of Transport and his officials to discuss parking policy, appropriate fines, and tourist coaches in Westminster; also giving evidence to a Parliamentary committee.

The engineering included the preparation of the design for an immersed tube tunnel under the Thames east of London; a project which eventually foundered when the proposed ringways around London were abandoned. We accomplished hydraulic studies and dredging tests and I hope the knowledge we gained and recorded was picked up by others doing similar work.

On the maintenance front the Albert Bridge gave us particular concern; nearly a hundred years old, it was designed by Ordish as a stayed girder bridge, but its sliding joints were arthritic and members which should have been in tension were buckled. There was no obvious strengthening which could be added so we decided to put in a central prop, closing the bridge meanwhile. As environmental groups objected to the idea of a prop a public inquiry was called, and I spent a day or more giving evidence. To the question of how often were Thames' bridges inspected I was proud to answer 'every day'; an inspector looked at each to see if there was a problem such as boat collision. The prop went in and the public in general did not seem to mind.

Peter extended his meeting arrangements to ensure good communication. Twice a week we met first thing for 15 minutes, working to a typed agenda delivered at five the evening before; the work demanded it and his department was efficient.

After seven years, at the age of 45, I was given the opportunity to become a partner in a firm of consulting engineers, something I had always thought I would like to do. You do not get many such chances at that age, and I decided to change; a lawyer succeeded me. I felt that we had accomplished a good deal in London: there was a dual carriageway route – part motorway – from Shooters Hill to Hackney Wick; Westway was constructed and there were other improvements; the majority of the capital's traffic signals were computer controlled, improving the average speed of flow; bus lanes were in use and Oxford Street about to be closed to cars; and road safety improved. Other world capitals were looking to us for an example of how to do things.

I was fortunate to be in the thick of urban highway and traffic development at a time when attitudes were changing fast, especially in the environmental field. Without being cynical, the wishes of all cannot be met in this latter respect, and the best that can be done is to spread the dissatisfaction as evenly as possible. Going to speak to people always proved fruitful, though; they appreciated that you were doing something.

Looking back, it was the most formative period of my career. The engineering content was good, but in addition I had been given the opportunity to meet and work with many people in various disciplines and walks of life, and to respect them: the police, civil servants, architects and planners, as well as politicians at local and government level. Having to promote and validate proposed schemes was new to me. I like to think I glimpsed the obvious occasionally.

Dudley Dennington

BSc(Eng), FREng, FCGI, FICE, FIStrucE, FIHT

Institution of Civil Engineers, member of council 1975–1978 and 1981–1984, vice-president 1990–1992. Greater London Traffic Executive, co-chairman 1970–1972. Greater London Road Safety Unit, founder chairman 1971–1972. Marshal Committee on Highway Maintenance, member; visiting professor of civil engineering, King's College, London, 1979–1982

Mr Dennington graduated in 1947 and after national service in the Royal Engineers joined Sandford Fawcett & Partners where he was engaged in the design of water-related projects. Following experience with contractors, D & C and Wm Press, supervising oil refinery construction, he joined George Wimpey & Company in 1952, and in 1958 was appointed manager of their design department.

Mr Dennington was appointed as assistant chief engineer highway engineer to the Greater London Council in 1965, chief engineer in 1967 and traffic commissioner and director of development in 1970. He assumed all executive duties arising from the Council's role as a planning, traffic and highway authority; during this period he implemented London's computerized traffic signal control system, brought in the Council's bus lane programme, and was responsible for the first urban motorways in London.

In 1972 he joined Bullen and Partners as a partner, where his responsibilities included major highway projects in the United Kingdom, and work in the Middle East and Hong Kong. He was senior partner from 1989 until retiring at the end of 1992, when he became a consultant to the firm. In 1994 he received an independent appointment from the Department of Transport in respect of contractual matters for a major road project.

He served twice on the council of the Institution of Civil Engineers and was vice-president for engineering in 1990. He has been involved with the Engineering Council since its start, and has been concerned with the training of young engineers throughout his career.

203

Learning to be a consulting engineer

Stefan Tietz

Faced with the question 'what were the formative issues in your career?', one realizes that some which are retrospectively important seemed less so at the time. Typically my school had a strong science side but my sixth-form education was on the arts side, a term which included everything but the sciences. This made for a tough start to my university education, trying to understand mathematical concepts which most of my year had learnt at school. My success in this was not initially conspicuous, more so as the allure of the social life and sports was considerable. The pay-back came much later, when I was asked to prepare reports on interesting themes in preference to able engineers perhaps less used to self-expression on paper.

Gaining experience

Early employment on the railways, with consultants and then with a contractor, provided breadth of experience but also made me recognize that I had a less than average tolerance for routine, repetitive work and bureaucratic procedures. This caused me to accept employment as the first and only technical assistant with what was then a tiny firm of consulting engineers, L.L. Kenchington, later Kenchington Little & Partners. An unanticipated bonus was that Lawrence Kenchington lectured two days a week, inevitably therefore leaving me in temporary charge though accountable for my decisions on his return to the office. A second bonus was that the firm grew. This gave me a very early insight into the workings of a practice, a very wide spread of technical experience and early interaction with fellow professionals and clients.

I suspect that my learning curve demanded considerable patience from Lawrence Kenchington but fortunately he enjoyed teaching. As the firm grew, responsibilities widened and again good fortune assisted as in addition to design I was interested in aspects of the work which others found less exciting, including data collection and an ambition to develop systems – often unsuccessfully. Typically I attempted to rationalize the many options available for precast and *in situ* flooring systems, by preparing tables comparing their design parameters such as amounts of reinforcement, shuttering etc., measured against span and cost. I learned much about the advantages of different floors but the aim of the exercise was never achieved as contractors' own preferences depended on their capabilities and their options as purchasers in the marketplace. The pipe dream of a rational basis of choice was therefore never realized.

An early surprise was the relative absence of problems which required higher mathematics, perhaps a disappointment bearing in mind the effort put into acquiring the relevant knowledge. Even at that early stage in my career it became clear that a logical approach, an interest in building beyond the narrow confines of the skeleton and in methods of construction was more likely to be relevant to most projects and also allowed one to solve problems outside one's specialist expertise.

Setting up in practice

Above all, this broader range of problems brought confidence and perhaps over-confidence as it resulted in my setting up in private practice at the age of 31. In a profession largely made up of consulting engineers with longer experience, this was perhaps regarded as slightly outrageous. As I worked totally on my own for the first year it also demanded attention to some obvious shortcomings. I lacked a sounding board for my ideas and therefore spent time within the learned institutions and particularly the Institution of Structural Engineers, first on their graduate students section and later on other institutional activities including two spells on the Council. Though it took up scarce time, it gave an invaluable insight into the workings of the profession. I also occasionally lectured at the Architectural Association and juried, their name for the assessment of students' design projects by staff, visiting specialists and the whole class. Their studio-based system and management of teaching, largely conducted by practising architects, was a revelation to anybody brought up in the normal engineering system where lecturers had often started as brilliant students, were encouraged to take a higher degree and were then retained for research and lecturing, therefore often lacking practical experience. Probably I learnt more from trying to teach architects about structures than they ever learnt from me, not least as it caused me critically to examine a vast range of architectural concepts, some of them highly original and complex, and also obliged me to jettison engineering jargon in order to make the subject comprehensible. This was equally valuable when discussing projects with clients and non-engineers.

Another gap in my education was practice and financial management. Again luck played a part as the course which I chose was clearly aimed at much larger practices. This obliged me to examine suggested management procedures more critically in order to adapt them to my particular needs. It taught me far more than I would have gained from a directly applicable course, where apparently adequate course notes would probably have led to a superficial acceptance at the expense of real understanding. Another fortuitous choice, consciously made but with little realization of the true long-term benefits, was my decision to live in central London. It greatly reduced my travel time and thus released time for a range of excellent evening courses and other benefits of a major city, many unrelated to work. Later, when my practice was bigger and required more management, it also left more time for design work. Through being close by, I was occasionally also asked by the media or public bodies to address some technical issue.

Early significant projects

So much for early formative experiences. Equally important ones grew out of the work undertaken. Though my practice grew slowly it did so consistently, to a peak of aboutt 70, barring the usual cyclic downturns which the industry has always had to accept. One very informative experience was the appointment to act as consultant to a large contractor. The brief was vague and for a retainer I was available as necessary to address the design impact of construction options, unexpected site findings or occasionally alleged construction defects. Increasingly I was also asked to examine invitations to tender to see whether some tweak to the design would produce an answer more likely to find favour with the potential client, in which case an alternative design might be put forward. Normally such options related to a better layout, easier constructability or economy. Occasionally this resulted in a new appointment as the consulting engineer on a development, but the problems were usually interesting even where no further appointment resulted. Frequently an appraisal had to be carried out within a day or at most a week – an excellent way of concentrating the mind as any proposals put forward had self-evidently to stand up to close later scrutiny if the option was accepted. It has always amazed me how these brief, concentrated exercises could result in designs which, when later substantiated by months of calculation and detailing, rarely differed by more than a few per cent in the final quantity of materials used or in a different methodology. I am firmly persuaded that most projects are now heavily over-designed.

Two projects, both very urgent, warrant further comment. The first was a large 'design and build' complex for Rolls Royce Composite Materials. They had won a contract for a new aero-engine incorporating novel components and the programme demanded that the first major phase, consisting of an industrial processing plant together with its ancillary facilities, a research block and offices, should be completed within eight months, including all design. The 30 000 sq metre complex included deep press pits, substantial overhead services, cranes in part of the building and other complications. The site was on a flood plain and had to be raised. All buildings including their ground slab had to be piled. The first challenge was thus to rationalize the design to achieve maximum repetition, using off-site assembly where possible. The second was to design out complexities. Thus framing for the flat roof construction over the crane bays was changed from that

INDUSTRIAL DEVELOPMENT FOR ROLLS ROYCE, AVONMOUTH.

adopted elsewhere in order to maintain the same roof levels and permit continuous cladding. A system was evolved jointly with the contractor for forming conical pile caps in the fill, lining these and then casting the pile cap and flat slab construction integrally. Reinforcement mats specific to the contract were designed and pre-welded off site for pile caps and bottom and top slab reinforcement so that the mats could simply be dropped into place. The reinforcement saved more than compensated for the cost of pre-welding. A system of tubular twin lattices was designed for the primary steel beams in the roof, fabricated off site by a nearby steel contractor. These sat on stanchions with a tee capping beam, contained many of the overhead service runs and provided access walkways at 16 m centres. The secondary tubular lattice girders, which also resisted out of balance stanchion loading, could then be simply supported onto these relatively stiff primaries, with lateral stability against stress reversals due to wind provided by the wiring and lighting tracts.

Another valuable lesson on this contract was the control procedure for maintaining the programme. Rolls Royce had not finalized their own designs and plant requirements. Procedures thus needed to allow for modifications. A very rigid updating system was adopted, whereby those changes which did not cause delay to the construction contract were adopted, while any others could only be adopted if a written acceptance of a delay was authorized. As only delays which did not affect the overall programme were acceptable, any others were postponed, to be dealt with through a supplementary contract after completion of the main contract. Interestingly, by the time this supplementary contract was undertaken, many of the changes earlier thought vital had not only become less vital but had often cancelled each other out so that most modifications finally made were minor.

Rather different points of interest arose on another project also required with great urgency. An explosion had demolished much of the client's factory. The expiry date of their consequential loss insurance meant that the new factory had to be fully operational within that period. We were invited to be the lead consultants, co-ordinating other professional input, including a range of specialist and complex services.

The original brief proposed a precise replica of the old factory. My analysis of the processes and the movement and storage areas for materials, however, suggested that historic additions made over time to the old factory had resulted in an irrational layout. The client accepted proposals for a total re-design as long as it

ROLLS ROYCE COMPOSITE MATERIALS FACTORY. *Factory block during construction.*

permitted the same construction and commissioning programme. To save time, the client also accepted a contract using approximate quantities, with costs based on schedules of rates and an overall budget estimate. In the end the final costs were within 3% of the estimate and the programme was met.

In order to reduce the client's stock of spare parts we asked all services providers to adopt similar hanger systems and other standardized components. While this aim was very welcome to the client, we underestimated the problems involved in persuading services specialists to adopt uniform systems. They tended to leave much of this more routine aspect of their work to their foremen and therefore did not know in advance what would be used. This obliged us to define the nature and location of all services with the suppliers, then draw all of these up on master plans, devise the necessary support systems and then discuss these with the suppliers' site management. It also demanded great patience from our engineers as all the details had to be fully drawn, a significant change from the normal ad hoc procedures with experienced foremen determining on site what should be done without drawings and with the first on site bagging the easiest runs.

The lessons learned firstly showed how much can be achieved in a short time if the procedures adopted also give priority to urgency. Secondly, they highlighted the problems of becoming involved in parts of the industry whose practices were different and, bearing in mind the undoubted benefits to the client, perhaps also indicated future scope for improved procedures when dealing with services. A measure of the benefits of the redesign was the very much greater production achieved in the new factory, compared to its predecessor of similar size.

Work as an expert

I used to have regular informal debates, often on traffic planning, with Leslie Ginsberg, a widely-known planning expert often appointed to public inquiries. This resulted in his recommending me as the civil and traffic engineering expert to give evidence to the public inquiry for the Winchester Bypass. The Department's plan was for the road to run through Winchester's famous water meadows, which raised heavy local objections. I believed that a new motorway was necessary though I also thought that the old road could be upgraded to urban motorway standards. In my response to the clients' brief I offered this, but my proposal was rejected in favour of one from another consultant which aimed to prove that the existing road would suffice. Subsequently the barrister representing the appellants advised that these proposals would not stand up at the inquiry. At very short notice we were therefore invited back, to develop the brief which we had originally proposed. This fascinating exercise demanded input from specialist experts, local residents and others with knowledge of plant life, local history, archaeology, the geology of the local limestone and many other factors. Traffic statistics and forecasts had to be critically examined and various options for the exact line of the road had to be examined to ensure that it could pass between the adjoining canal and St Catherines Hill, of archaeological interest and thus to be preserved. Several types of retaining wall and some bridges had to be designed and costed in sufficient detail to ensure their viability, all of this in a limited time.

As a result of a preliminary decision by the inspector, the local road construction unit was asked to co-operate with us to draw up our proposals in more detail and then to present our joint findings to the inquiry. This too proved a stimulating exercise, also for the civil servants who were used to a less open-ended way of working. An interesting facet of such inquiries is the number of new issues constantly arising from new evidence, often needing rapid appraisal to brief lawyers for the following day's cross-examination.

The inquiry also underlined the importance of being personally involved in the details when preparing evidence. When cross-examined I was given a far easier time than the county engineer, responsible for many more projects and a much larger staff and therefore inevitably less conversant with the details prepared by his staff for the inquiry.

Would I do the same, starting today?

In the last ten years circumstances within the industry have changed dramatically, not least through the widespread application of computers to design, management and drafting. The processes are also more bureaucratic, the industry is more litigious, firms tend to be larger and designers are given less freedom to develop original ideas. Nevertheless I still believe that one can have just as interesting and diverse a career, with the chance to work in many countries, given the will to work over a sufficiently broad field and accept some risks which in my opinion are an inevitable by-product of undertaking original design. Engineers' career structure aims at developing the expertise necessary to exercise judgement and it is this proper exercising of judgement which should be of greatest value to our clients. That does not however remove all risk!

Stefan B.Tietz

BSc Eng, FEng, FICE, FIStructE

Stefan Tietz is a consulting civil, structural and traffic engineer and founded S.B. Tietz and Partners in 1959. He is now a consultant to the firm. He gained early experience on railway bridges, then with consulting engineers and contractors, and now works extensively as an expert in disputes. He also advises on projects to agree the design brief with the client, carry out preliminary engineering appraisals or reappraisals to achieve design improvements or savings.

Stefan Tietz has also worked extensively abroad, projects including the appraisal and design of ski resorts. He is a member of several committees and working parties at the Institution of Structural Engineers, is a director of the Timber Research and Development Association (TRADA) and chairman of its Research Advisory Committee.

Loads that can lose their way

Malcolm Woolley

The construction industry by its very nature is a hazardous business. Health and safety regulations and the introduction of the planning supervisor have gone a long way to ameliorate the inherent dangers of the process. Many accidents occur through human error and are related to the behaviour of site staff as they go about their daily operations. The cluster of structural failures, some with very dramatic and horrendous consequences that are the subject of this essay, were accidents of a different type. It was a lesson for the whole civil engineering industry and the consequences have remained with us to this day in the form of bridge checking procedures and what are known as the Merrison rules. The headline events all occurred within a relatively short period in the late 1960s to early 1970s in the UK and elsewhere. It is the story of a development phase in bridge design and construction, although the issues at stake and the lessons learnt have the widest possible application in structural engineering.

There was much excitement in the air at the time, with new materials, new ways of using them, and new roads, bridges and buildings to be built in profusion. Engineers were talking across national boundaries about structural analysis methods and their application to practical problems of the day. Designers felt that they were at the forefront of a struggle to meet society's needs and provide value for money. That need still remains today.

Let's start at the beginning: structural efficiency, by which is meant the economical use of the appropriate material to be used in a particular design, is the holy grail of the structural engineer, so the very idea that a structural element can do two jobs at the same time is most attractive. In the immediate post-war years with rationing and material shortages a recent fact of life, minimalization was the order of the day.

For the avoidance of doubt, I should perhaps confirm that minimalization means what it implies, that is, to use minimum material to achieve a set objective. It is not quite the same as structural efficiency, but closely related. The principle of minimalization holds good for today, but then it was a singular aim, with formal whole-life costing and value engineering yet to be recognized.

FIGURE 1 *Hammersmith Flyover, London designed by the Maunsell Group in the 1960s. An early example of concrete box beam construction. Note the increased depth and, therefore, stiffness adjacent to the piers.*

The struggle between the two primary construction materials, steel and concrete, pushed minimalization to its limit. Concrete had the apparent advantage of relatively low basic cost with the added attraction that it could be moulded into any (almost any) desired shape or profile. This property was being exploited either in a precast factory environment or cast *in situ* with the use of temporary falsework. Concrete structures can remain heavy however in comparison with their steel counterparts, and this drawback enhanced the struggle for structural efficiency, thinness of sections, etc.

In parallel with concrete, the development of steel bridge structures was largely the development of welding technology, the use of thin plates and the associated fatigue issues. The cluster of significant structural failures in the 1960s–70s was evenly populated with steel and concrete as the basic structural material.

The words stress and strain are in the popular mind inextricable from any consideration of engineering materials, although perhaps they are only vaguely understood and often confused with load and movement.

The higher the stress, that is the load per unit area, the more likely there is to be failure. A lot of design effort, therefore, goes into ensuring that stresses are within acceptable limits. Strains, on the other hand, are a measure of the amount of movement that a particular stress induces (change in length/per unit length). A stiff structure, one in which there is relatively little movement under working load, has many apparent benefits over its flexible counterpart. The one disbenefit of a stiffer structure, that in being stiffer and stronger it can actually attract more load

213

FIGURE 2 *Second Severn rossing approach viaducts built by Laing GTM, completed in 1996: the most recent and dramatic use of concrete box beam construction in the UK.*

than its flexible counterpart, was to be a deciding factor in the box beam failure cluster.

The search for structural stiffness was a component (but not the only component) in the popularity of continuous, rather than simply supported, structures. In broad terms continuity makes multi-span structures about five times stiffer than their simply supported counterparts. Similarly, the two-way spanning slab, particularly when contained by edge stiffening, has a stiffness which is of a different order of magnitude to its one-way spanning, simply supported, counterpart.

Continuity has the other benefit of adding an element of redundancy into the structure, which in turn allows alternative load paths to develop before ultimate load is reached. Redundancy is widely used in nature in order to promote survival. Plants produce many more seeds than will finally germinate; the human body has two of most things and can manage in many cases with only one. Similarly redundancy as applied to structural design is generally welcomed, especially if it is available at no cost to structural efficiency and element minimalization.

As with all things, there is a downside to structures that contain redundancy. It means that loads have a choice in which way they are carried, and it is the responsibility of the designer to determine the factors controlling the load path and cover for the alternatives. It is here that *loads can lose their way*, and this point will be returned to later in this essay.

Figure 3 *Multispan bridge: simply supported deflection profile.*

Figure 4 *Multispan bridge: continuous deflection profile. About five times stiffer than the simply supported counterpart.*

All this is common knowledge and only repeated here for the sake of filling in a little of the background of how in the 1960s we came to be designing new forms of structure to fulfil well-established needs.

So much for structural efficiency and redundancy. There is another major all pervading issue for all prefabricated items, that of tolerance or *lack of fit*. Here is an issue that spans manufacture and erection. Steel as used for structures has that vital property, yield, and how we take advantage of it. Lack of fit is most often overcome by force with the local steel obligingly yielding to accommodate the construction tolerances.

In the case of concrete, the situation is more delicate and we need to introduce an *in situ* element in the form of packing and grout, or more specifically an *in situ* stitch, to make sure of a good fit. It is as well to be reminded that in the generality, to allow for the relative brittleness of concrete and to ensure that we have due warning of failure, we codify ultimate carrying capacity, so that our steel reinforcement yields and the concrete cracks in tension long before a brittle concrete compressive failure occurs.

The design opportunity of the time was to mobilize the benefits of torsional stiffness (or twisting strength) offered by the box cross-section as a bridge spine beam replacing the then more conventional multi-beam and slab construction for bridge decks. The consequence was the potential to create an unfamiliar distribution of load at points of support, particularly during the construction phase, for reasons which I will explain (see Figures 5 to 8).

With the benefit of hindsight it is easy to see how the box beam bridge idea became adopted worldwide as a symbol of good engineering design and good value for money (an expression then yet to be coined for popular usage), some in concrete and some in steel. In order to get the flavour of the time established in the mind, it is important to remember that then structural design was carried out primarily using a slide-rule and/or seven figure logarithm tables with just the possibility of the laborious use of a mechanical/electrical calculator.

Given these conditions and the use of the Morice & Little bridge deck analysis based on moment distribution and slope deflection methods of the time, you would allow about six weeks for number-crunching the critical load cases of a typical bridge deck. Similar input and output today might take a matter of a few hours or be hidden away within the software such that an 'off the shelf' bridge or structure is produced for many of the standard or near-standard cases required.

Certainly in planning the design process today the situation is very different, most time being consumed in client and design team interaction with the analysis process being largely taken for granted. This is where I believe *loads can still lose their way* and is the central point of this essay, because the advent of computers and computing power has not altered the fundamental need to understand what is going on and why structures behave in a certain way – how in particular *added stiffness can attract added load* and at points in a structure where there is *a change in stiffness there will always be stress hotspots*.

By stress hotspots I mean uneven or non-linear stress distributions which will need special design consideration. The simplest example which comes to mind is when a rectangular hole is cut in a web or slab. Theory indicates infinite stress in the re-entrant (internal) angle of the hole. As an aside, although the first jet airliner, the Comet, was a huge success, the prototypes were dogged with problems caused by high stress concentrations around the fuselage windows: a stress hotspot caused by a change in cross-section.

Returning to the main theme of this essay: accidents are always the result of a conjunction of events, each of which if acting alone would not have had the same out-turn. It is always a matter of judgement as to which singular factor was the primary cause of failure, and which was secondary or tertiary. Commonly there may be as many as three or more readily identifiable contributions to a structural failure. In the case of the cluster of bridge failures in the 1960s, the common factor if not the primary cause, was the presence of the box beam form, whether in steel or concrete.

The box beam has immense torsional (twisting) stiffness when compared with other forms of grillage or beam and slab structure. To shape your primary load-carrying element in the form of a box was to kill two birds with one stone. When combined with deck cantilevers, any out-of-balance live loading was transmitted to the pier 'in one' via the torsional stiffness of the spine beam boxes already in place to carry the main longitudinal loads. Mancunian Way and Hammersmith Flyover are two of the first, and excellent examples of this form of construction in the UK. Both were designed by Maunsell Consulting Engineers and constructed without major incident.

It was the very high torsional stiffness of the box, so attractive in its finally constructed form, that was to embarrass so many projects in the construction phase. Three-legged stools, like three-wheeler cars have a limited application; in structural terms they are simply supported. With wide beams we welcome that fourth support, two at both ends, and feel that we have a stable arrangement with an element of redundancy to cater for out-of-balance loads. Witness, for example, the double tee form of construction widely used for floor slabs in precast concrete. In the case of the box beam, often with slender webs not necessarily positioned directly over the bearings, highly stressed in the construction phase, the ever present torsional stiffness of the main box beam combined with adverse construction tolerances could very quickly place a disproportionate load into one web. This had the potential consequences of placing all the load on the one bearing instead of two, simply doubling the load (Figure 8).

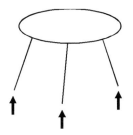

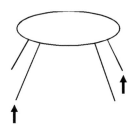

FIGURE 5 *Whatever the lengths of the legs of the three-legged stool, loads are evenly spread although resistance to overturning is limited.*

FIGURE 6 *Leg lengths of the four-legged stool are critical. With only one leg shorter than the rest, the majority of load can soon be concentrated in two legs only.*

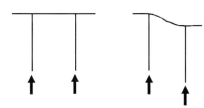

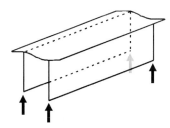

FIGURE 7 *A beam grillage is torsionally flexible and accommodates construction tolerances through this flexibility.*

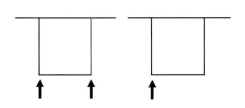

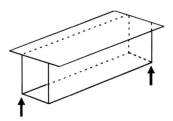

FIGURE 8 *The torsion box, because of its torsional stiffness, can tend towards being supported predominantly on two supports only if there are any inaccuracies in construction (diametrically opposite corners each end).*

FIGURE 9 *River Camel Viaduct designed by Gifford and Partners. Note the simple lines, constant depth of construction and repetitive spans. An award winner from the 1990s.*

With construction tolerances alone capable of creating such significant over-load conditions, there needed only to be one other adverse factor to create a critical condition; for example an apparently minor support settlement or (as in one case) the use of mild instead of high-strength steel in the temporary works.

Each generation faces the issues before it with a different and ever-developing perspective. In the case of designers, in the current climate there is the fear that their importance to society is not understood and is, therefore, undervalued; a feeling that, with the advent of very powerful computers and the attendant com-munication revolution, all we need to do is talk and the computers will do the rest.

Through all human development there remain common threads. In the case of the structural designer the fight to adequately resist natural forces remains constant and very real. Society's centre of interest may have moved from monu-ment building towards communication but, in so doing, it takes the safety of its structures for granted. We may be able to talk up for example the value of a particular share on the stock market, but we are unable to talk down the forces and the distribution of forces acting upon our structures. We can only analyse through a thorough understanding of all relevant issues, especially the practical construction issues involved.

In this the computer is a great help, but left to itself is rather dumb. Rubbish in – rubbish out, and all that. One cannot emphasize too highly the importance of the 'what if' scenario; the need for the sensitivity analysis not only for the costing aspects of the work but in order to ensure that *'loads do not lose their way'* as they appeared to do in the cluster of failures in the 1960s–70s.

In summary then, I remain enthusiastic for the continuing search for structural efficiency and new ideas using both traditional and new materials. I was privileged to have been personally associated with a particular development, the box beam bridge, now firmly established in the hierarchy of bridge design options with numerous examples around the world successfully completed and functioning well. This career experience caused me to move from the particular to the general, with some broad conclusions that remain valid independent of technological developments in the design process that have intervened.

These are:

- How *added structural stiffness can attract added load* with significant consequences which may be beneficial or otherwise; the converse being how structural flexibility can spread load often in a helpful way.
- The overriding consequences of *lack of fit* in the construction process and the importance of sensitivity analyses. What if?
- Where there is a discontinuity or change in section, *stress hot spots* are bound to occur due to change in stiffness, and these need close individual attention.

The computer is well able to deal with all these issues, but it does need the knowing eye in concentrating its analytical powers in the right areas and interpreting the output. In other words, the design function remains pre-eminent in determining the starting point and judging the finishing point, the computer notwithstanding.

Now that market forces rule and competitive design in a design and build situation is more often the norm, there remains great scope for a renaissance in structural design in order to add significant value to the customer or client. In the same way that the stock market can easily out-perform the deposit account, allowing for one or two failures along the way, so the bespoke design produced by the right designer will always out-perform the 'off the shelf' standard.

The message has to be – keep the computer and communication methods in their rightful place – in the service and not in control of the procurement method. Allow individual judgement to surface in competition and let the best man win. *Loads should not lose their way in the hands of designers who have not lost theirs.*

References

1 Calder Bridge M1 motorway, Yorkshire. Gifford et al. *Proceedings Institution of Civil Engineers 1969*, **43**, August, pp. 527–52.
2 Cleddan Bridge, Milford Haven. *Highway and Road Construction*, May 1975, **43** no. 1785, London.
3 West Gate Bridge, Koblenz Bridge et al. *New Zealand Engineering*, November 1974, **29** no. 11, Wellington.

Malcolm Woolley

BSc, CEng, FREng, FICE, FIStrucE, FIHT

Malcolm Woolley was educated at Bishop Vesey's Grammar School, Sutton Coldfield, before graduating from Birmingham University with 1st Class Honours in Civil Engineering in 1953. Between 1953 and 1959 he worked with specialist contractors designing and supervising the construction of reinforced and prestressed concrete structures, and within this period obtained a postgraduate diploma from Leeds University in concrete technology.

Employment with Gifford& Partners started in 1959 with a mixed bag of design commissions involving prestressed and reinforced concrete buildings and bridges and many timber structures including hyperbolic parabolic shell roofs.

Winning the Ministry of Transport design competition for a bridge to carry the M1 across the River Calder in Yorkshire in 1962 was a career milestone. A partnership in Gifford & Partners followed in 1971, Managing Partner in 1979, Senior Partner in 1986 and Chairman in 1996. He now acts as a consultant to the practice with which he has been involved for nearly 40 years. The practice is best known for its achievements in bridge and building structure designs and has many awards to its credit. Activity is centred in the UK with a developing overseas component involving all types of commissions ranging from archaeological through to major buildings and major road and rail schemes.

If you want God to smile tell Him your plans

James Armstrong

Few of us are men and women with a world-shaking destiny. We like to think, with hindsight, that we played a part in shaping our lives, but the most that can be said is that we are beset constantly by the most improbable coincidences, of which we may or may not be aware at the time, and which we may or may not grasp and turn to our advantage. The world, and its human population, form a rich and exciting stage upon which we play our parts, and a talented and supportive cast with whom it is a privilege to work. In looking back on the play, with its many delightful plots and sub-plots, we sometimes feel that we have influenced affairs, but it is rather like watching one's grandchildren make sand-castles – a busy decision-making process, entirely directed by the available resources and of a transitory nature!

But there seems to be some purpose in life; being given by birth human talents and curiosities, there seems to be an implicit duty, a responsibility, to use these talents, to get to know more of the creation and its laws, and to encourage others to do likewise – what Socrates called 'divine curiosity'.

I am grateful to all those who encouraged my curiosity, who guided me in this human task, and helped to develop the skills needed to understand a little and, above all, to enjoy life, contributing, it is hoped, a little to improving the quality of the lives of others in the process.

Childhood influences

Rearranging the elements to meet needs seems to have been characteristic of my family. I never knew my paternal grandfather, who died whilst my father was still a boy, but both my father and my maternal grandfather, James Hodgson, were of a very practical turn of mind. Father was a carpenter by trade, but moved into the motor world, building commercial vehicle bodies onto bare chassis, and developing a workshop full of fascinating machines for shaping the timbers, where I spent many happy hours exploring the possibilities of the variously shaped off-cuts – using equipment which I am sure, in these more safety-conscious days, would be out-of-bounds to eight-year-olds! Father's success in business enabled me to become the first undergraduate in our family's, probably turbulent, Border history. He had a high reputation for his acumen and for his integrity, truth and compassion. He became chairman of the local Chamber of Commerce.

My grandfather was a builder and engineer, and unsuccessful inventor (he had two business failures). But he was employed as some sort of part-time technician/ consultant by the firm of Cowans Sheldon, internationally renowned as crane manufacturers and based in Carlisle, my home city. He designed, among other things, winches. One of my earliest recollections, probably in 1934/35, is being driven in the open 'boot-seat' of his little Ford to West Cumbria, to Maryport where he was to advise on the dredging of the harbour, using a static winch to haul a drag line bucket to and fro across the harbour.

I have a clear memory of standing, hands seriously thrust into trouser pockets, listening to the deliberative, very forthright and very Cumbrian, discussions on the merits and demerits of various tactics that might be employed. It seemed a good thing to be doing!

No doubt there were many similar incidents, but the only other clear recollection of my grandfather's influence was after his final retirement, when he used to help with the family garden. I was returning from University in 1945 for a short break, and met him in the garden. He had a dandelion in full seed in his hand, which he gave to me, saying, '...and they say there is no such thing as a God.' It confirmed the mystery, and the need to work to understand and to apply the understanding; it was the least that one could do. That was the last exchange we had; he died shortly before I finished at University.

School years

Our value systems and interests are forged not only by family examples, but by our peers and our teachers. The Grammar School in Carlisle, where I was educated from the age of eight until I went up to Glasgow to University at the age of 17 in 1943, has a good reputation, and provided several influential teachers. An early interest in the delight of geometry burst suddenly, under the guidance of a junior schoolteacher, a Miss Gamble, into the joys of abstracting the hidden order in the world of the visible as concepts and relationships in the world of the intelligible, the art of proof.

A spinning bowl in the physics laboratory, containing a viscous fluid, part of an experiment by a very serious sixth-form boy, caught my attention. The beauty of the surface form – a paraboloid – was fascinating, and the master in charge took the time to explain to the lowly fourth former how it could be described mathematically.

Whether these details are symptomatic of a love of order or whether they themselves induce that love I am not sure. I suspect that a knowledge of order is inherent in all of us, and these events bring it out into the open, and the circumstances of our lives enable us to take advantage of them in particular ways. Perhaps other circumstances would have produced a musical response, or an artistic result, or a literary ability – but in me they seemed to result in the wish to apply my enthusiasms to the physical world in which I found myself.

An important asset in the realization of engineering projects is the ability to communicate. I was fortunate therefore, in having an English teacher who inspired a love of literature, a precision in formulation, a respect for syntax, that have been of great value. In particular I recall a great joy in the art of the précis! Reducing statements to their essentials seemed good! Perhaps this leads to efficient and economical designs. A favourite quotation of mine is from Newton's *Principia*: 'Never use more when less will suffice, for Nature is pleased with simplicity and affects not the pomp of superfluous causes.'

University

During the 1939–45 war the choice of universities was very limited, and few were offering civil engineering courses. I was fortunate in securing a place at Glasgow, where a general engineering course was offered, including significant elements of mechanical and electrical engineering as well as civil engineering subjects.

We were particularly fortunate in having as a senior member of the staff Hugh Sutherland, who has remained a close friend and mentor throughout my career. Hugh had worked with Karl Terzaghi in the field of soil mechanics, and the elegance of his applications of mathematics and basic physics to such apparently unpredictable materials as rock and soil I found quite breathtaking. Hugh had a quietly persistent and totally rational way of explaining what was being discussed; love without excess emotion! It set a high ideal in the practice of engineering, meeting human needs using the wisdom of the order in creation. This was further reinforced by studies of the elegance of Maxwell's laws of thermodynamics and Southwell's work on relaxation techniques, the precursor to finite element analysis.

One of the subjects we studied – in classes which began at 0800 – was Higher Natural Philosophy The tradition was maintained of the distinctions between moral and natural philosophy in Glasgow at that time, now, regrettably, no longer the case. I cannot confirm that the abstract knowledge of the shape of a trajectory in a non-isotropic viscous fluid has been applied by me in practice since that time. However I can confirm that the fact that such practical phenomena could be modelled using, if I remember correctly, fifth order differential equations, impressed me immensely. It has given me confidence in the power of analysis and mathematical modelling which I have found of great use since. It was reminiscent of the statement by Henry Palmer in, I believe, his inaugural address as President of the Institution of Civil Engineers on the 2nd January 1818: 'The Engineer is a Mediator between the Philosopher and the Working Mechanic, and like an interpreter between two foreigners, must understand the language of both, hence the absolute necessity of possessing both practical and theoretical knowledge.'

Because of the intensity of the degree course at Glasgow during the war, the four-year 'thin sandwich' programme of the traditional Scottish degree at Glasgow was condensed, by taking out most of the industrial training periods, and I graduated just before my 20th birthday in May 1946. I was undoubtedly well trained, but, with hindsight, not very well educated, compared with the greater maturity of today's graduates.

Practice

The background of education and training has been set out in some detail above because it seems to me an important precursor of the specific projects upon which I have been engaged at various points during my career, and was probably more influential than the apparently more important events later in my career.

I have taken as illustrations two aspects of my enjoyment as significant – the application of the laws of nature to physical problems, and the encouragement of others to give of their talents as members of multidisciplinary project groups – the natural philosophy of things and the moral philosophy of human desires and aspirations.

Natural philosophy

There are three examples of analysis which I recall as particularly satisfying from differing periods in my career, which seem to embody the work of the engineer as interpreter between philosopher and working mechanic. Each served in its own way to confirm the intrinsic delight in order, and the extrinsic satisfaction of using this delight to serve some identified need.

Elliptic shell roof (1948)

Working for a small Edinburgh consultant, David Kerr Duff, I was asked to design an elliptical concrete shell roof for a paper-mill extension. An approximate model was created, involving the use of an assumed analogous fine net, and the distribution of loads and stresses around this net. It involved learning something of elliptic integrals, and the manual relaxation of successive elements, using a modification of some of Southwell's work. There were no computer resources available at the time, and the analysis took, if I remember correctly, some six to eight weeks of painstaking work. The discipline developed during my years with Hugh Sutherland proved invaluable. But I do not remember being bored at any time; watching the orderly arrangement of the forces emerging as the design developed, coupled with the feel for structural behaviour that seemed to have a life and intelligence of its own, yielding to and accommodating the loads, was more than adequate reward for the work involved!

Soil-structure interaction, British Guiana (1958)

Using experience gained on site with buoyant foundations for the oil refinery at Grangemouth, a design was developed for a floating cellular raft to support a bulk sugar warehouse in British Guiana. The raft was about 300 ft long by about 120 ft wide, and about 12 ft deep. It was sitting on a very soft estuarine silt subsoil more than 100 ft thick. The maximum loading was about 20 000 tons, fluctuating considerably throughout the year. The problem was to judge the deformation pattern of the raft, in order to arrive at the likely distribution of loads and stresses and hence suitable reinforcing and prestressing figures.

I had the great pleasure of working with Dr Robert Gibson, then at Imperial College. He is an acknowledged world leader in the nature and behaviour of soft soils, and in their mathematical modelling. Bob Gibson worked on the three-dimensional settlement possibilities, assuming a variety of possible soil structure reaction patterns, and I calculated the deformation patterns of the structure using the same reaction patterns, and Timoshenko's theories of the behaviour of slabs on elastic foundations. We plotted the results of these movements of structure and soil until we arrived at a match between the structural deformations and the soil deformations. Acknowledging that the natural behaviour of the structure and soil in their interaction would be at least as intelligent as we, we assumed that the final deflected forms would also be compatible. We then used the interaction pattern given by this principle of superimposition as the pattern to be accepted in reinforcing the structure. It is still working. Our confidence in the intelligence of the natural materials seems to have been justified! Bob is still a close friend!

Channel Tunnel traffic flows (1986)

The third example of natural philosophy at work arose during considerations of the analysis of traffic flows at the Channel Tunnel terminal works at Folkestone. We were faced with the problem of coping with a maximum design capacity of vehicles at the terminal of about 3000 per hour. Their arrival pattern was random, but their departure pattern on the shuttle trains had to be measured and orderly. There were about 17 different types of vehicles and vehicle combinations (cycles, coaches, cars, caravans, HGV, etc.),all driven by independent, nervous and potentially erratic individuals!

Standing one day on an overbridge in Cannon Street station, I found myself looking down upon individuals disembarking from a crowded suburban train, and vanishing down a flight of stairs to an underground walkway. Each of the passengers was confident in his or her existence as an individual decision-maker, but viewed from above their movement pattern was one of typical turbulent flow, with eddies of counter movement at the discontinuity between platform and staircase similar to those of markers in an hydraulic testing tank. The idea of using a hydraulic model was adopted as the basis for the construction of a mathematical model. It was also recognized that the insertion of filters and 'porous' screens into the flow (taking the form of toll booths, customs and security checkpoints etc.) would convert the turbulent flow of the random arrival pattern into a more streamlined departure flow. This included the insertion of a 'stilling basin' into the system prior to boarding, and diversion channels to enable traffic to visit the facilities buildings, with every driver believing he or she was personally making the key decisions ! Once again there was the satisfaction of sharing in the natural intelligence of Nature's laws!

Moral philosophy

But engineering is about people, about their needs and abilities, their use of natural resources, their interactions with one another. The mathematics and physics of the elements and forces in nature must relate to human behaviour, willingness and resistance, wisdom and ignorance.

Site work

Perhaps the introduction to engineering as a human activity begins with the opportunity to relate to the working mechanic on construction sites. As a junior resident engineer on the Scottish Oils refinery at Grangemouth I was one person within a labour force of perhaps 1500 – all more experienced than I! The site foreman was known as Big Tim Tynan, for good reasons. He was a larger than life general foreman with Wimpey, but very considerate of the needs and vulnerability of young engineers, with a surprising respect for instructions given from such a naive source! I learnt from him the respect that must be given to size and weight, a quiet confidence that, if treated with respect, it was possible to lift gently and locate accurately fractionating columns weighing 120 tons, to locate and sink to within an inch or two of the required depth concrete caissons weighing thousands of tons, and to care for the well-being and safety of the workforce in the process.

During my time on site there was a pit disaster in Ayrshire. I turned up on site one morning to discover that nearly all the labour force and most of the heavy plant had vanished overnight to work on the pit, sinking a shaft to locate some of the buried miners. When your life depends upon the care taken by your friends, you learn the essence of teamwork. Men who, in their general behaviour, seemed to be all toughness and hard living, become delicate and caring, with a fine accuracy in their movements. Watching steel erectors or piling riggers at work was like watching a ballet, every movement attended to with full attention. There is clearly a close parallel between the delicate elegance of applying refined mathematics to the solution of a problem in the abstract world, and the elegance of touch needed to make the work visible and available in the material world, with the added moral connection of care for the well-being of others. On site your neighbour is more evident than he is in the design office and it is easier 'to love him as yourself!'

Peer influence

Moving back to a design office in London, after four years on site in Scotland and the north of England, I found myself in the company of three or four other young engineers working for Rendel, Palmer & Tritton, under the overall guidance of John Cuerel, who had been responsible for the design of the new Waterloo Bridge over the Thames. I remember them with great affection: John Hutchinson, who later became Professor of Soil Mechanics and Engineering Geology at Imperial College, Jeffery Johnson, who went to Grenoble to work with the hydraulics research unit, and John Dennis who remained at RPT, later to become a partner.

We formed our own study group, which met at lunch-times, to discuss very seriously what we were interested in or learning from our work. I can recall a series of visits to City churches led by John Hutchinson, a discussion I led myself on electro-osmosis (which we had used at Grangemouth), and papers by Jeffrey on hydraulics. This activity must have continued for about a year, and was invaluable in learning to appreciate the interdependence of professionals upon one another for their personal as well as for general professional development. It was a useful introduction for the more extensive work with which I became involved later, and which still continues, with the professional institutions.

BDP and George Grenfell Baines

A very significant move took place in 1963. I had been developing my career as a specialist in soil mechanics, perhaps following Hugh Sutherland's example and enthusiasms, but felt that I needed to make a change in order to achieve a more responsible position. I discussed the options with an older, respected, barrister friend who advised me to look for a large organization dealing with larger complex projects, rather than becoming too much of a narrow specialist. My attention had been drawn to the development of the Building Design Partnership, guided by the architect George Grenfell Baines (GG). I made enquiries, attended a conference at which he was a key speaker, and was impressed by what I heard. I wrote to him offering my services! He declined the offer, but six months later wrote inviting me to apply for a position in his newly opened London office. I joined the organization, staffed by young and enthusiastic architects and engineers, and began work on hospitals, universities, shopping centres, office buildings, etc.

This introduction to collaborative team working, with an acknowledgement of the interdependence of the several members of the construction team, considerably expanded my interests. Involvement in design sessions with planners, architects, service engineers, quantity surveyors, landscape architects and later also with interior designers and graphic designers provided a completely different scale of enjoyment. GG himself was a charismatic leader, encouraging, listening, questioning, but never domineering. His leadership was inspirational, and he always gave credit to all members of the team. It has been a privilege to have him as friend and mentor.

As a partner with BDP I have had the opportunity to lead several major project teams, with horizons widening throughout my career, so that I can honestly say that life has continued to become richer and wider, even after retirement the impetus has continued through educational and institutional activities. Much of this I attribute to the refusal of GG ever to accept any limitation; failures were always opportunities to learn and to improve the contribution to the next project.

Institutional work

I began to attend meetings at both the Institution of Civil Engineers and the Institution of Structural Engineers shortly after my arrival in London in 1954. By some mysterious process I found myself invited to join working parties and committees at both institutions. Most were concerned with engineering education and training. I suspect the influence of Hugh Sutherland at the Civils, and certainly the extension of my activities at the Structurals owed something to suggestions from Peter Dunican.

My interest in education and training has continued to this day, and has been a source of great satisfaction. There are few, if any, better ways to consolidate one's experience than to pass it on to others. There are many to whom I am indebted for this interest and opportunity, too many to present in this essay. But the knowledge of being part of a continuum of service to the community, acquiring, refining, formulating and communicating aspects of the engineering tradition and collected wisdom, continues to be one of the most rewarding aspects of my career.

I owe a deep debt of gratitude to all those who have guided me. I might mention in particular David Simms of Kingston Polytechnic, now Kingston University, Professor Jean Le Mee of the Cooper Union College in New York, Dr Frankie Todd of Leeds University and Adrian Long of Queen's Belfast, all of whom gave me the opportunity, as a Visiting Professor, to participate more fully in the educational process, and to enjoy the company of future generations of engineers.

Paradoxically, the move into high office at the institutions, whilst a great honour and opportunity, seems less vivid and earth-shaking than some of the earlier experiences, although I learnt more of my fellow engineers' contribution to the profession and to society during my year as President of the Institution of Structural Engineers than I had in the whole of my previous career! Perhaps one's youth is more vivid and character forming than the tasks and services undertaken as a responsibility in maturity. Seniority brings with it a stabilizing role, less dramatic, but vital in providing measure and good judgement to one's chosen profession.

In recent years, there are two projects which have been very much concerned with people. The Royal Academy of Engineering's scheme for the placement of Visiting Professors in the principles of engineering design, whose steering committee I have chaired for several years, has provided the opportunity to visit many universities and to take part in the planning and delivery of engineering education, for all engineering disciplines. Through this work I have made many new friends, and am learning a great deal about current developments and future opportunities and responsibilities for the engineering industry, sustainability, globalization, systems engineering, etc.

The second project has been contributing to the redrafting of the Rules of Practice of the Institution of Civil Engineers, intended to bring them into line with current developments in society, environmental issues, consultation and participation, contractual procedures, etc. This focus on ethical issues has brought together the delight and responsibilities of engineering with another powerful area of experience which has proceeded in parallel with my practice as an engineer, the study of Philosophy.

Philosophy and engineering

I left university in 1946 satisfied that I had been well trained, but would have welcomed the opportunity for a fuller education. After a few years of training in practice and becoming a chartered member of the institutions, I listed the books that I felt I should have studied, but had not. They included the Bible, Shakespeare, Plato, Bertrand Russell and others! On arriving in London I saw an advertisement for a course on practical philosophy, and joined the evening classes offered by the School of Economic Science in 1955, in Suffolk Street, off the Haymarket in central London. These philosophy classes had grown out of the study of economics, it being considered that you could not study the nature of society without a knowledge of the nature of mankind. I have been a member and tutor at the school ever since.

The method of study was by a form of Socratic discussion, which I, and twelve months later my wife, found very satisfactory, allowing plenty of opportunity for exploration. The focus was on the discovery of one's own nature rather than on the accumulation of historical or 'scholarly' information.

The Senior Tutor in the school was a barrister, Leon Maclaren, whose father, Andrew Maclaren, had been a highly individual MP for many years, with a particular interest in the economics of land ownership and use. The introduction to the ideas of the world's great philosophical traditions, eastern as well as western, has been of immeasurable help in bringing together as a whole all the diversity of both current and past events in the history of man.

These studies have led to the development of short courses on reason, ethics and aesthetics that I tutor from time to time. Evolving the material for these courses has moved in parallel with my professional practice, and has proved time and again to be very relevant to the process of analysis and synthesis that is design, the relationships between the various parties to any major project which are largely ethical, and the wish to participate in the production of elegant constructions, which is aesthetics.

These wider interests have provided the opportunity to lecture around the world. In the last six months I have been to Australia and South Africa lecturing at universities and to other groups on a variety of subjects ranging from the Channel Tunnel to Plato's four Virtues – wisdom, justice, temperance and courage!

Little, if any, of this would have been possible had I not been ably and continuously supported by Marjorie, my wife of almost fifty years. Her tolerance and encouragement and her sharing in the delights, sometimes involving sacrifices on her part, have been essential in achieving what I have found to be a wholly satisfying and enjoyable career.

James Hodgson Armstrong OBE

HonDEng., FREng., FICE, FIStructE, FASCE, FRSA

Dr James Armstrong is a member of several professional bodies, and Past President of the Institution of Structural Engineers.

He has played an active role in monitoring and developing educational policies in professional education. He has played a significant part in co-ordinating and developing the work of the Royal Academy of Engineering Visiting Professors and chairs the Academy committee considering design matters in engineering. He has been a visiting professor and lecturer at several Universities in the UK and elsewhere, and is currently Visiting Professor at Kingston University.

His career has been predominantly as a consulting engineer. He retired in 1989 as partner in the Building Design Partnership, where he was responsible for planning and designing such major projects as the Channel Tunnel terminal works, the Falkland Islands Airport, and the University of Surrey.

He is particularly interested in the development of design abilities, and in multidisciplinary design team projects. He lectures on design and design management both to engineers and to the general public. In 1996 he contributed to the redrafting of the Institution of Civil Engineers Code of Professional Practice.

He is chairman of several charities, the Higher Education Foundation, the European Christian Industrial Movement, the Varanasi Education Trust, and the Maryport Heritage Trust. He is a governor of the Independent Education Association and the four associated St. James' days schools. He was Chairman of the Harris-Manchester College, Oxford, building committee. He is a member and one-time treasurer of the executive committee of the School of Economic Science.

If you don't put a date to something you can't even be wrong

Lawrance Hurst

*I*f only someone had impressed on me years ago the importance of trying to date buildings and bits of buildings and persuaded me to read the book setting out the dates when various materials came into and went out of general use, I would be so much further ahead now. Unfortunately, no-one did point out that importance and so far as I know the book has yet to be written, so I am now only just starting to learn about and appreciate this vitally important subject.

Dating buildings and bits of buildings is the best way to learn about how building construction developed and the history of any particular building. If you fail to think about the date of a building or a bit of it, you have lost the opportunity to be wrong and you will never learn. If you try this mental exercise and spend a few minutes checking how correct you are, you will either have the satisfaction of knowing you were correct or gain some knowledge because you will actually know you were wrong. If you do not even try, you will remain ignorant.

It is all too easy to look round an existing building and to jump to conclusions about its condition or about the answers to questions put to you, without really seeing or thinking about the construction. If you consciously look at the components that go to make up the external elevations, and each room you visit and think hard about when they were made or when they were put together and which came first, you will understand them better. If you have an idea of the date of construction, your knowledge of the development of types of construction will give you some idea of what you can expect to find. If you find something else, which perhaps you would regard as coming from a different period, you suspect an alteration or an extension, or perhaps that form of construction or type of floor, or those bricks or that sort of window, or that pattern of plaster was used at an earlier date than you had thought.

Charles Brooking, at the University of Greenwich, who has been collecting building components for many years, has made a study of buildings and can date a window from the sash pulleys. At the age of three, Charles Brooking began to collect house numbers. By the age of fourteen, he was collecting doors, windows and ironmongery, and had declared his intention to found a professional reference centre. That commitment has not changed. Today his dream is a reality and the Brooking Collection is housed as a living museum at the University of Green-wich. His particular interest and expertise is in windows of all sorts, particularly timber and metal, which he can date and source with remarkable accuracy. He is always interested to hear of examples, frequently from buildings being demol-ished, which he can add to his collection and his knowledge.

Some would regard such information as irrelevant or too esoteric, but the development of a general knowledge and feeling of dates can provide enormous satisfaction and also be of very real value when considering or working on existing buildings.

Readers who dismiss this recommendation because their work is all on brand new buildings do themselves a disservice. Few buildings are on new greenfield sites or do not adjoin or come close to existing buildings. A building to be demol-ished to make way for a new development needs consideration and understand-ing for at least two reasons. Demolition needs an understanding of the building and its structure if it is to proceed smoothly, without the sort of discoveries that can result in accidents or even premature collapse. Gaining that understanding is simplified by consideration of the date of it and each piece. When the timber skirtings are removed from an upper floor room to reveal Fletton bricks laid in cement mortar, you know, or you should know, that the Georgian terrace house you thought you were looking at has been extended or altered. But you can only know this if you are aware that the fields at Fletton were sold for brickmaking in 1877,[1] and you will not find pressed bricks fired from Oxford clay before that date, and furthermore it is unusual to find brickwork in Portland cement mortar in ordinary buildings before about 1900.[2]

Your existing building to be demolished is probably also next to other build-ings which have to remain. This gives you another opportunity to develop your knowledge of dates. Which came first and when, what sort of structure would you expect and can you see any signs of alterations or of surviving earlier construction which could affect the implications of your demolition for the adjoining buildings. Method statements have become essential for most construction processes and when written for demolition they are all about avoiding surprises. If you know or suspect that old loose fragile construction may be immured in newer construc-tion, you are able to suggest exploratory work to discover whether you are correct, and hence to avoid the surprise of a partial collapse which would have resulted from the disturbance of the status quo by your demolition. If you do not think about the construction and its dates, you are at risk of being surprised.

It is fairly general knowledge that 18th and 19th century London stock facing brickwork is commonly not well bonded to the ordinary brickwork of which the rest of the wall thickness is composed.[3] If you know or discover the date of the

[1] Hillier, Richard. *Clay that burns – a history of the Fletton brick industry.* LBC, London, 1981.
[2] Hurst, B.L. Concrete and the structural use of cements in England before 1890. *Proc. ICE,* **116**, Issue 3 & 4, Aug/Nov 1996.
[3] The lack of bond between facing and backing brickwork in the 18th and 19th centuries is referred to by many authors, both contemporary and modern and throughout the intervening years. It is regularly re-discovered!

A TERRACE OF BUILDINGS IN FARRINGDON ROAD.
These buildings are Victorian from their style and appearance but Farringdon Road is a 'new' street opened in 1845–6, renamed in 1863* but not developed until the 1880s, so they are probably later than that.

*Weinreb, B. and Hibbert, C. 1993. *The London Encyclopaedia*, revised edition, Papermac, London.

CLOSER EXAMINATION REVEALS THE DATE OVER THE GATEWAY...

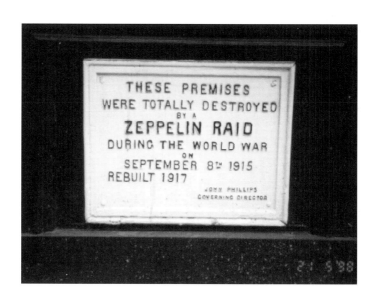

THESE PREMISES
WERE TOTALLY DESTROYED
BY A
ZEPPELIN RAID
DURING THE WORLD WAR
ON
SEPTEMBER 8TH 1915
REBUILT 1917

JOHN PHILLIPS
GOVERNING DIRECTOR

...BUT IT WOULD BE WISE TO EXPECT THE MATERIALS AND
CONSTRUCTION IN THIS ONE TO DIFFER FROM THE REST.

building, you are able to make use of this knowledge to avoid problems when altering the building or demolishing next door. If you do not think about the date, you may proceed in ignorance towards problems or even a collapse, which is not an accident because it could and should have been foreseen.

Most engineers, and indeed other building professionals, unconsciously develop some feeling for dates and also unconsciously make use of it – the modern kitchen units or flush panelling in a half-timbered thatched roof cottage indicate alterations have been made and hence some mutilation of timbers and internal alterations would be expected. If however this feeling for and thinking about dates is conscious and is developed, it will increase the interest of the work and should be of inestimable value in the carrying out of work.

If you are looking at or round a building with someone else, a discussion of the dates of this and that eases thinking and helps reasoning – four eyes can thus see much more than two times two. If you are alone, consciously discuss and question the dates of this and that with yourself. If you do it consciously then you have to think about the dates and thinking about the dates makes you really look at this and that, not just see them. Dates are a very useful peg on which to hang your thoughts and deductions about all parts and components of an existing building.

Another useful mental exercise is to discuss with yourself the dates of the buildings you pass as you go about your daily business. This makes you look at them and their details, and also makes you appreciate the architects who have incorporated the date in their design so that you can know the satisfaction of being correct or have the benefit of knowing you were wrong and thinking why. If you fail to look at them and date them you will pass by in ignorance, and not even know that the information and experience you could have gained would be useful to you next week.

Sometimes, and particularly to start with, it takes courage to date buildings or parts of buildings or components, because you simply do not know and are frightened or ashamed to be wrong, even to yourself. It is well worth striving to overcome this mental block and to gain the habit of dating buildings as you pass them and of parts of buildings or rooms you enter or pass through. It will pay dividends.

When I was at school there was a saying that talking to yourself was the first sign of madness, but it is also said that madness is akin to genius, and if only I had realized at the start of my professional career the importance of dates, and had started then to apply my mind to dating, I would surely by now have been a genius. I would also have positively recorded the advent of new materials and noted others going out of use and would not have to ask colleagues to rack their brains with me to remember when hollow clay tiles for floors ceased production, or when we started to put polythene under ground slabs — and I would be able to write the book mentioned in my first paragraph.

Where to find information for dating

Buildings (style)
- Books on styles and architectural details
- Date incorporated in elevation or on foundation stone
- Old maps, particularly large scale OS maps
- *Pevsner's Buildings of England*
- Local books e.g. *Survey of London*
- Local reference libraries
- Contemporary publications e.g. *The Builder*
- Post Office directories published annually since about 1850
- Rate books

Construction and components
- Old building construction books
- Books on old or historic building construction
- Contemporary specialist publications
- Libraries e.g. RIBA, ICE, IStructE
- Ask a specialist – e.g. Charles Brooking for windows

Lawrance Hurst

BSc, ACGI, CEng, FICE, FIStructE, FConsE, FBEng

Lawrance Hurst was educated at Oundle School, graduated at Imperial College, London and served in the airfield construction branch of the Royal Air Force for his National Service, before joining Andrews Kent & Stone in 1957. After five years as an Associate at AKS he moved to Hurst, Peirce & Malcolm in 1968 where he was senior partner and is now a consultant.

Hurst, Peirce & Malcolm, founded by his father in 1910, is a small practice of consulting engineers for structural work connected with buildings, new and old, in most materials.

Lawrance prefers old buildings and he has been fortunate to have been concerned with the Royal Albert Hall, the Nelson Monument, the London Custom House and Finsbury Barracks, as well as a number of other less prominent but equally important and interesting old buildings. He has written papers on 19th century concrete and cements, fireproof flooring, iron and steel sections, E.O.Sachs, and the history of party wall legislation.

His appointments to advise on structural aspects of work to party walls, in which he has become something of a specialist, give him the opportunity to see a large number of existing buildings and to practise dating them and their components.

A leaf blown by the wind

Lord Howie of Troon

My working life has not been an orderly one with planned steps along a logical path, but has had more than a hint of randomness about it. In a career which has embraced civil engineering, politics, journalism and publishing, it is clear that there has been more than one influential event or turning-point. Indeed, there have been several.

The first of these, which pointed me toward civil engineering, was educational. The small town of Troon on the Ayrshire coast, where I was born in 1924 and brought up, had no secondary school then. The most it had to offer was the Higher Grade School, known affectionately as the Big School, an elementary establishment, the pupils of which would either go on to Ayr or Kilmarnock Academies or, more likely, leave altogether at 14 years of age. Since my family had no tradition of advanced education, it was fairly likely that my education would come to an abrupt end.

Fate came to my aid in the unlikely shape of Charles Kerr Marr, a local industrialist who had amassed a substantial fortune. Luckily, he was also a philanthropist and he left a large sum which he wished to be devoted to local education. After some legal wrangling, it was decided that this money should be used to build a new secondary school in Troon and provide bursaries to local students in higher education.

The school was opened in 1935 as a co-educational comprehensive, which was common in Scotland, and as a direct grant, which was not. Being well funded, it attracted first-rate teachers and rapidly gained a reputation for excellence.

This could hardly have happened at a better time for me, as I passed the Scottish equivalent of the 11-plus the year the school opened and entered it in the following year. Not relishing the curriculum of the top stream, which stressed French and Latin as its main studies, I opted for the second stream which was founded on French and technical subjects. Thus, I became acquainted early on with mathematics, mechanics and technical drawing, three of the staples of an engineering education.

But I was not yet set on the engineering profession, of which I then knew nothing, nor had the thought of higher education entered my head. The next influential event came when I reached the age of 15 or so and was considering leaving school. Having shown some aptitude for engineering drawing, I applied for a junior job in the shipyard which was then Troon's main industry. Since most of my family had worked there as riveters, ships' carpenters or on other ship building trades, this seemed a logical enough progression. To my surprise, being unversed in such matters, I was informed that the office took on nobody lacking the Higher Leaving Certificate, which is the university entrance qualification in Scotland.

This apparent setback was anything but, for it meant my staying on for my Highers. In passing, I may say that although the Highers do not aspire to quite the 'golden standard' claimed by the English 'A' levels, they provide broader education. In my case, the core subjects were buttressed by physics, chemistry, botany, English, history and geography. I have never found the lack of 'A' levels to be a handicap in any of my activities.

My Highers behind me, I noticed that many of my contemporaries were going on to university or other higher education institutions, encouraged by the prospect of Marr bursaries as well as natural ability, and detected a minor twitch of ambition which hinted that I might follow their example. My education clearly pointed me toward some kind of engineering, but I had no idea to which kind I was suited. Careers advice at school, which was quite primitive in those days, was of no real help and I had lost the notion, which had never been a compelling one, of shipbuilding.

Troon came to my aid. It had, and still has, a splendid harbour and was the seaward terminus of the Kilmarnock & Troon railway, the first proper railway in Scotland, which was prominent in local folklore. Both had been engineered by William Jessop, of whom I had never heard, and were significant works of construction. Furthermore, in local government terms, Troon was what was then known as a small borough with its town council and an enterprising one at that. So it had its own programme of public works including a water supply with a reservoir at Loch Bradan in the South Ayrshire hills.

I was aware of the borough engineer's work and thought it to be socially useful, though he had also designed the local swimming pool, an edifice of small architectural merit. So, my choice was made and I followed him into his profession with a vague aim of public service, and enrolled at the Royal Technical College, Glasgow, now Strathclyde University, in September 1941 as an unworldly 17-year-old. Three years later, I had graduated and was ready for the world of work. If this appears to be a haphazard approach to the choice of a career, all I can say is that it was.

For the next eight years, I was employed in Scotland, partly as a structural steel designer with the Colville Construction Company in Mossend and later with Babtie, Shaw & Morton, a consultancy which I remember with abiding affection and with whom I became a corporate member of the Institution of Civil Engineers. Thereafter, I joined the drift south and spent the next 11 years in London, mainly with Rendel, Palmer & Tritton (RPT). By then, I had moved into more general civil work and, by chance, became a minor expert in prestressed concrete design.

How that came about is the next influential event in an unstructured career. At my interview with RPT, I was asked to outline my experience, as a guide to my usefulness. As I explained, I had designed heavy steelwork, river and harbour works, sundry contributions to hydro-electric schemes and several reinforced concrete structures of various kinds. The interview seemed to be going well until I was asked if I had done any prestressed concrete. Not only had I never done any, since this was the early 1950s, I had scarcely heard of it. White of face, I confessed. Luckily, my ignorance did not stand in the way of my appointment, and I got ready to join this famous consultancy.

On arriving on the chosen day, I reported to the senior engineer who had interviewed me, ready to receive my instructions. I was assigned, together with a colleague, Peter Cox, later a President of ICE, to the design of a bridge in Takoradi Harbour in what was then the Gold Coast. The fly in the ointment was that Peter was to handle the approach spans in reinforced concrete while I was to design the modest prestressed concrete main span over some railway tracks.

Taken somewhat aback, I reminded my superior of my ignorance of this technique and hinted that this might not be the best use of my limited talents. My objections were swept aside with an instruction to swot the subject up as quickly as possible. This I managed to do, and in the next ten years I designed numerous prestressed concrete structures culminating in my being one of the design team on the prestressed pressure vessel for Wylfa Nuclear Power Station, possibly the first in Britain. I could hardly claim to be a prestressing pioneer, but these were early days, in this country at any rate.

My ten years with RPT brought my career as a civil engineering designer to an end, though as it turned out not my connection with the profession. But then came an influential event which turned my life in quite another direction. I had been brought up in a political family with both parents active members of the Labour party. So I was exposed to politics from an early age and developed an abiding interest in political affairs and theory. It was not long before I turned to practise at first in Scotland and then in London and while still with RPT I fought an unsuccessful parliamentary election in 1959. My colleagues were well aware of my activities at that time and perhaps saw them as a quaint, even bizarre, deviation from the civil engineering norm.

At any rate, shortly afterwards I was asked to stand at the subsequent election at Luton which was regarded as a winnable marginal seat. The influential event was a cabinet reshuffle which removed the sitting member, Dr Charles Hill, from office and which led in due course to a by-election in late 1963. This was an event of some significance since Labour's capturing the seat with a decent majority was regarded in political circles as a pointer to the general election which was expected a year later. So it turned out and Harold Wilson came to power with me as a minor cog in the government machine in the Whips' office.

My House of Commons career was not lengthy, lasting for seven years until I was swept away in 1970 in Ted Heath's somewhat unexpected victory. While Luton could not be expected to withstand the adverse swing which was obviously coming, it seemed that Wilson's government would survive, though damaged. That was not to be, however, and Downing Street changed hands. Though saddened by Wilson's defeat, I was not downcast by my own. I had enjoyed the managerial nature of the Whips' office and afterwards the more political activities of the Council of Europe and backbench select committees. Moreover, while in the Commons, I had had three years on the ICE Council and had tried, with no great success, to bring the Institution closer to the six chartered engineers in the Commons, two each from the big three institutions.

After all that excitement, I was delighted to be asked back to RPT, no longer as a designer, for the profession had gone metric without me, but to look after contracts and such esoteric matters as settling claims and final accounts. When I am asked if I found politics to be on the dubious side from time to time, I reply that it has much in common with settling a claim: thoroughly enjoyable, if you happen to be that kind of person.

But, much as I liked to be back in the profession, I had changed. Parliamentary life is demanding but, since it lacks the usual disciplines of office life, it has an illusion of freedom about it. The nine to five routine seemed more shackling than the 16-hour days ever did and, after three years, I needed a change.

It came in the form of yet another influential event and a complete break which, in fact, gathered together the threads of my political and engineering experience and gave me a new, third, career. The Civils, thanks largely to the drive and foresight of Garth Watson, then ICE secretary, had established its weekly magazine, *New Civil Engineer*, in an attempt to reach out to its members and give them news of the industry and profession which they might otherwise not get. This was a daring initiative since news leads to opinions and, sometimes, opinions lead to criticisms.

Despite some initial nervousness on the part of the Institution's establishment, fanned from time to time by journalistic brashness on the part of the magazine, it has to be said, this experiment has been a successful one. To a great extent, this was due to the shelter which Watson gave to the paper's journalists when they became overly enthusiastic.

At all events, Sydney Lenssen, then *NCE* editor, engaged me as public affairs correspondent with a very wide remit and, in November 1973, I became a full-time journalist in a niche which might have been tailor-made. Nearly 25 years later, I am still in that niche, though reduced as far as *NCE* is concerned, to a monthly column.

At first, the prospect of a former Labour MP commenting on an industry which is not unduly radical in its outlook and responding to the policies of a Conservative government raised a few eyebrows and there were some nail-biting moments. With time, suspicions were allayed, perhaps because I was able to demonstrate reasonable, if not total, objectivity.

Even then, the turning-points of my varied career had not ended. For, in early 1978, not quite five years after I had become a journalist and while I was a member of the committee of inquiry into the engineering profession chaired by Sir

Monty Finniston, I was made a Life Peer at the instigation of the then Prime Minister, James Callaghan.

In my 20 years of active membership of the House of Lords, as what is nowadays known as a 'working Peer', I have come to respect the Upper House and the work that it does, especially as a revising chamber and in its select committees. As a revising chamber, it is far superior to the Commons and does that work much more thoroughly.

The bull point in the Lords' favour is that its members are not primarily politicians, though they do have political views and allegiances, or at any rate many of them do. More often than in the Commons, they speak from a standpoint of professional experience and expertise. For that reason, it is possible for a backbench peer to get amendments to legislation on the strength of the fact that he might well know what he is talking about better than the Minister does.

Looking back, I can see a random life, marked by distinct strokes of luck. Now, as a Life Peer and a journalist with civil engineering tying my three lives together, I could probably give the impression of careful planning and forethought. But that would be wrong. Influential events which proved to be turning-points are the clues to my career. A leaf, blown by the wind, might say as much.

Lord Howie of Troon

CEng, FICE

Born in 1924, Lord Howie was educated at Marr College, Troon, and the Royal Technical College, Glasgow (now Strathclyde University).

Currently a director of PMS Publications Ltd and other companies, Lord Howie originally trained and practised as a civil engineer. From 1963 to 1970, he was Member of Parliament (Lab) for Luton and held a number of posts in the Government Whips' Office before becoming a Vice-Chairman of the Parliamentary Labour Party.

Shortly after leaving Parliament, he became public affairs correspondent on the *New Civil Engineer* magazine, and he is still a regular correspondent on that and other magazines. In addition, Will Howie was for a time a trade union adviser to the Institution of Civil Engineers, and he has published two short books on trade unionism with particular reference to professional engineers and the construction industry.

Lord Howie was, from 1984 to 1991, Pro-Chancellor of the City University. He is a former President of the Association of Supervisory and Executive Engineers, the Association for Education and Training Technology and the Independent Publishers' Guild. At present, he is a vice-president of the Periodical Publishers Association.

In the House of Lords, he is a member of the Select Committee on Science and Technology and has been a member of the European Communities Select Committee. He is also a Member of the Societé des Ingenieurs et Scientific de France and holds honorary doctorates from City and Strathclyde Universities and is an Hon. Fellow of the Institution of Structural Engineers. Lord Howie was a member of the Finniston Committee of Inquiry into the engineering profession. He has numerous publications to his name.

An enormously satisfying career

Brian Simpson

Looking back over nearly 50 years of structural engineering has made me reflect on why it has been so satisfying a career. My thoughts have gone back to those early days when every job seemed a new challenge and was in a way a puzzle to solve. When I was first introduced to loading codes and to BS 449:1948 as a teenaged apprentice working in a steel fabricator's design office it seemed to me that, once the mysteries of shear force and bending moment were understood, designing steel frames for buildings would be straightforward, and a job for life. As my apprenticeship progressed I found that there was more to learn. By the time I was 21 my self-confidence was such that I felt I knew as much as the next chap even if I did not know it all. This was put to the test and generally knocked out of me during the next two years of National Service, a period which many regarded as a career break and a waste of time. I had been encouraged by a colleague who had just returned from his two years' National Service absence from designing steel frames to apply for a commission and was fortunate to be awarded one. During training, I was introduced to many aspects of engineering new to me such as Bailey Bridging and demolition by explosives as well as aspects of military engineering like minefield laying and clearing.

This was in the days when as a student designer one was expected to provide all one's tools of the trade: a boxed set of drawing instruments and, the pride of every budding designer, a slide-rule with as many scales as you could afford, even if you seldom needed more than the basic A, B, C and D scales. One essential aspect of slide-rule calculation was to have an approximate idea of what the result of a calculation would be, principally to be able to put the decimal point of the answer in the right place. This process did develop a certain agility in mental arithmetic, which served me well when I was learning to be a sapper officer during my period of National Service. Cadet training took the form of strenuous physical activity in the mornings followed by attendance at lectures in the afternoons. During one such lecture we were being told of the importance of undertaking simple calculation without the aid of slide-rules or logarithm tables. To test our ability in this difficult area, the instructor was picking on individuals to give the square of numbers he thought of. Seeing me dozing off he brought me awake by asking me for the square of 7½. I saved my bacon by giving him the answer 56¼ without hesitation. He then tried a few others with 5½ and 8½ with no success. 'Simpson and I know a trick that the rest of you don't know.' I may not have been popular with the rest of the gang, but this put me in favour with one of those who would decide my destiny at the end of the course, and I was grateful for having learnt the simple rule for squaring a number which is an integer and a half.

It arises from the formula

$$(A + B)^2 = A^2 + 2AB + B^2$$

if $B = \frac{1}{2}$, then

$$(A + \frac{1}{2}) = A^2 + (2 \times A \times \frac{1}{2}) + \frac{1}{2}^2$$
$$= A^2 + A + \frac{1}{4}$$
$$= A(A + 1) + \frac{1}{4}$$

so $$(7\frac{1}{2})^2 = (7 \times 8) + \frac{1}{4}$$

The extra ¼ is hardly relevant, but gives that touch of accuracy, which was evidently appreciated by my instructing officer at Chatham all those years ago.

The serious point of this story is that my generation was brought up with the concept of mentally checking our calculations so that gross errors were infrequent. Of course, errors could still occur, but the need to estimate the order of magnitude of our results was a healthy discipline, which does not seem to be widespread nowadays.

Having been taught the processes of sizing beams, columns, simple trusses and latticed girders, I was given the task early in my training of helping to determine the precamber of lattice girders in a complex roof over a bus garage. This was required because drainage of rainwater from the roof was channelled along valleys located above the top booms of the main latticed girders. It was necessary to compensate for dead load deflections to enable the surface water to flow along these valleys. This involved a procedure for determining the deflected shape of latticed girders known as a Williott-Mohr diagram using a graphical process to represent strain movements in each member of the frame. I did not meet this procedure again until it appeared in my course of evening class studies several years later, and I remember being grateful that I had been taught it early on in my practical training. I think I could still do Williott-Mohr diagrams if required because the process was so well explained and I had the opportunity of doing several examples 'for real' when first instructed.

Many of these design processes which we used before computers came to our aid must now seem esoteric, but their application gave an understanding of structural behaviour that I think must be more difficult for students to grasp now when so much analysis and design is automated by application of computer programs. Since so much is new in all aspects of our lives, we feel we must know more than our predecessors. They did not have the benefits of computer-aided design and modern codes of practice. Some of the materials they used were primitive by comparison with modern ones. Their understanding of the forces of nature was more limited than ours. We may not know it all, but we do know a lot more than they did. We may have better tools at our disposal, but our knowledge of materials and forces acting on our structures is only better because we have a better grasp of the uncertainties contained in our assumptions. They recognized the limitations of their assumptions and consequently adopted generous factors of safety.

Apart from the intellectual satisfaction of designing and problem solving, I see another element of my work which has contributed to my enjoyment of practising structural engineering.

As an apprentice working in the design office of a steel fabricating contractor with offices next to the stockyard and fabricating shop, I quickly assimilated what would now be called the culture of corporate identity. I learned which were the landmark projects of the past that were spoken of with pride. Dawnays had provided steelwork for such buildings as the Gaumont State cinema in Kilburn. I knew of this from old promotional material and hearing about it from old-timers who had worked on it. Providing the structural steel support for the balcony area without columns intruding into the stalls was stretching steelwork fabrication and erection technology to its limits at the time. As I cycled past the Gaumont State on my way home from work, it always registered – almost as if I had had a hand in it myself.

Then there was the greater thrill of seeing going out of the yard pieces of steel that I had actually sized by calculation! I had made my contribution; more than that, I felt complete responsibility for *my* steelwork. This feeling of exclusive rights to the product did not last long. I found that everyone in the firm who had designed, detailed, cut, drilled, welded, delivered and erected the steel, felt part of the project.

As my own experience widened, I came to realize that this personal identification with the work in hand, particularly on large buildings and structures, was part of that pride in personal achievement which gives so much satisfaction to all those engaged in the construction business. Virtually all construction work is teamwork where many trades and professions make their contribution to the final product and all take pride in the collective achievement.

I realize that identification with the product is not confined to structural engineering or to the construction industry. It is also wider than the pride in personal achievement. Seeing the work of others one has an appreciation of their skills in the physical presence of the objects they design and produce.

Going back to my apprenticeship, I recall the interest and the excitement of seeing photographs, reading articles about and actually visiting the Festival of Britain and seeing such landmarks as the Dome of Discovery and then the Skylon. I am still fascinated by innovation and the ability to design and to erect bigger, better and more elegant structures.

Brian Simpson OBE

CEng, FIStructE, FRSA

Brian Simpson was educated at Kilburn Grammar School and Brixton School of Building. His training and early experience in structural steel design was gained with Dawnays Ltd and a national service commission with the Royal Engineers.

He joined Husband and Company's London office in 1956, working on industrial structures and steel frames for radio telescopes, movable bridges, and roll-on/roll-off ferry terminals. He has been involved in the 'Bridgeguard' assessment of bridges, the Merrison assessment of steel box-girder bridges and bridge designs including the reconstruction of the Britannia Bridge over the Menai Strait.

In 1972, Brian was taken into the Husband and Company partnership and when the firm merged with the Mott MacDonald Group in 1990, he became a director until his retirement in 1992. From 1992 to 1998 he was a consultant.

Brian was a member of Council of the Institution of Structural Engineers from 1980–83, Vice-President 1991–4 and President 1995–6. He served on the BSI Drafting Committee for Steel, Concrete and Composite Bridges from 1985 to 1998 and on the Council of the Association of Consulting Engineers from 1987 to 1990.

Reflections on a desert island

Sir Jack Zunz

*T*he idea for the title of this piece came from the radio programme where invited guests have to discipline their musical predilections to eight, and no more than eight, choices. I thought it would be a challenge to choose eight projects, present or past, which for one reason or another have intrigued me more than others and to see if there is a common thread and whether any conclusions could be drawn. Those of you familiar with *Desert Island Discs* will know that the castaway is also allowed the Bible, the works of Shakespeare and one other luxury; we'll come to that later.

To make a list of projects which have been of interest, ones in which my colleagues or I have participated or ones which have been historically inspiring, has not only been difficult, but salutary. It has been like selecting what to put into your backpack for a long journey, where every daily convenience, be it a book, item of clothing or a pair of shoes becomes a necessity: how will I do without it?

I made a long list. It included Maillart's bridges, which I have admired since I first heard of them more than fifty years ago, much of Brunel's work, many of the great cathedrals, not to mention Paxton, Torroja and much more. Ultimately the selection process was impulsive rather than methodical, but I will try to give some reasons for my choice. I will also check, at the end, if there are identifiable coherent threads or whether it was simply a question of random indulgence.

Of the eight projects four are old – 100 years or more – while the other four are ones with which I have been involved to a greater or lesser degree. The order is neither preferential nor chronological; it just happened.

My first choice is the Forth Railway Bridge. The bridge's centenary was celebrated a few years ago and it was a timely reminder, particularly to the younger members of our profession, of its place in engineering history. Even by the exuberant standards of the Victorians, the design and construction of the Forth Bridge was a towering achievement. Fowler, Baker and Arrol have found their niche in the history books. Fortunately Stewart, whose analytical work has been recognized more recently, has also now been placed alongside these giants.

Remember that Sir Thomas Bouch's Tay Bridge had collapsed a few years earlier. Bouch, the original engineer, was then removed from the project. Despite the inevitable reaction after the Tay Bridge disaster, Fowler, Baker and their colleagues proceeded to design and construct one of the great bridges of all time. Remember too that steel was a relatively new material, at least new in the sense of large-scale commercial application. More than 50,000 tons were used with confi-

THE FORTH RAILWAY BRIDGE.

dence and panache. Analytical tools were fairly rudimentary by today's standards, yet the confidence was underpinned by exhaustive tests. The main struts in the bridge are fabricated 12 foot diameter tubes, more than 300 feet in length. The phrase 'less is more' was coined more than half a century later. Yet this bridge was designed on the principle that fewer members meant that fabrication, construction, erection and maintenance were simplified while at the same time making the bridge clearer, more comprehensible and less fussy. It is one of those rare structures which not only gives a feeling of confidence in its fitness for purpose, but is also visually commanding and satisfying. That it has stood for more than a century with minimal structural maintenance is a glowing tribute to its builders and designers.

My second choice is an unbuilt project. Many, if not most, unbuilt projects, particularly competition winners, contain untested optimism and fantasy conveniently forgotten when published or filed in the archives for posterity. They have yet to overcome the combined pressures and rigours of commercial reality and structural integrity. In this particular case most of the fantasies had been dealt with and the project was fully designed and detailed and was actually sent out to tender. I refer to an Indoor Athletics Stadium in Frankfurt, or the 'Leichtatletik Halle' as it is properly called. The scheme was the result of a limited architectural competition held in 1981. We joined Foster Associates in their entry and won the competition – always a satisfying experience – at least for the first euphoric twenty-four hours.

A PHOTOGRAPH OF THE MODEL OF THE NATIONAL ATHLETICS STADIUM, FRANKFURT.

The programme called for an indoor sports stadium with a 200-metre running track, sprinting tracks, seating for 3,000 people, and so on. The site is a football pitch surrounded by trees situated between the large grandstands of the main football stadium, the home of 'Eintracht Frankfurt', and a small sports hall, the Wintersportshalle. The concept, which is a fine example of the total integration of engineering and architecture, evolved around a sensitivity to the wooded site and a desire to create an unobtrusive low-key elegant envelope. The structure, mainly a vault based on the segment of a circle of 70-metre span, is partially sunk into the ground. The vault rises only 9 metres to give a total maximum floor to roof clearance of 15 metres. The shallow protrusion of the roof kept the site edges free for circulation and maintained a separation from the flanking buildings, while preserving views through the end glass walls on to the two wooded aspects of the site.

The plan dimensions of 140 × 70 metres militated against a two-way spanning arrangement and, while parallel trusses spanning 70 metres were considered, their frequent need for bracing suggested a two-way triangular grid, which also offered opportunities for integrating lighting, both artificial and natural, as well as the mechanical services, in an elegant and coherent manner. The structure, which after detailed design weighed no more than 35 kg/sq metre, was to be constructed wholly of tubular steel and was to form the language for the distinctive architecture of the building. It was highly repetitive, to be site-assembled of diamond-shaped units (see below). The low surface to internal volume helped the

CAD IMAGE OF THE REPETITIVE DIAMOND-SHAPED UNIT OF THE NATIONAL ATHLETICS STADIUM, FRANKFURT.

heat loads, the merging of roof and walls allowed more uniform daylighting on the arena, and the buried abutments could call on stiff reaction from the dense sand sub-soil.

The total integration of structure, heating and ventilation, roof lights and artificial lighting is seldom achieved. The choice of a slender and efficient structural frame was therefore quite indivisible from the choice of servicing, cladding, construction and aesthetic preference. This design is at the top of my list of unbuilt projects. It just might have become a fine leisure building. Its demise was due to a change of political colour in Frankfurt's local government and policies for providing sports facilities duly changed: a disappointment, albeit a democratic one.

My third structure is one which I selected from a large number of intriguing buildings made of unbaked earth. These are not the kind of structures which normally come within the ambit of the structural engineer, but they have had in the past, and could still have, a profound importance in building, particularly in third world countries. Some of the most interesting examples are to be found in West Africa.

The influence of Islam was superimposed on indigenous traditions and locally available building materials. Some quite ingenious solutions using only the building materials at hand, namely timber, vegetable fibres, earth and some stone, are found in the construction not only of houses, granaries and other domestic forms, but also for major public buildings, particularly mosques. Some of these mosques are more than 500 years old.

A fourteenth-century north African scholar (Ibn Khaldun) recognized the necessity of geometry for an understanding of the engineering problems associated with the construction of these buildings. The relationship between mathematics and the building sciences was explicitly acknowledged. Carpentry, it was suggested, 'requires either a general or specialized knowledge of proportion and measurement, in order to bring the forms from potentiality into reality and for the knowledge of proportions one must have recourse to the geometrician'.

Unfortunately information on much building of this period is speculative and some of the very oldest of these structures remain, thanks to their imaginative design. They are, however, little known. The grandest is the great mosque at Djenne in Mali[1], West Africa, near the southern edge of the Sahara (see page 254). Waterspouts on the roof ensure that rainwater from the roof does not run down and soften the mud-plastered, adobe-brick walls. The most fascinating feature of some mosques in particular, is the wooden poles which protrude from those walls (see Figure). These poles not only give these structures their distinctive architectural expression, but also provide permanent scaffolding for workmen to repair the surface of the building with a coat of mud-plaster. When it rains, which is of course fairly rare, some of the earth of the sun-baked facing is washed away. There is something very satisfying about the forms of these buildings as well as their relaxed relationship with their environment.

---

[1] *The Great Mosque at Djenne*. Raoul Snelden.

THE GREAT MOSQUE AT DJENNE.

One of the eight structures is bound to be Sydney Opera House (see opposite). Enough has been said and written about this building. A.J.P. Taylor said that 'history gets thicker as it approaches recent times', and there are some interesting aspects of the Opera House story which have not seen the light of day. This is not an appropriate place to tell them.

On the other hand, the Opera House has become one of the more significant buildings of the century. Whatever the debate about its final functional usefulness concludes, the original concept which fired the imagination of all who worked on it was extraordinary, it was a poetic vision which seemed tantalizingly within reach. Alas, human frailty (or was it an unrealizable dream?) intervened, but what there is now for Sydneysiders and visitors alike is pretty good by any standards. Debate about the functional adequacy of the interior planning will continue, but the structural integrity remains unimpaired, and its expression is the basis of the visual impact of the building.

I realize that its basic structural logic is open to debate. But there are few structures whose logic cannot ultimately be debated for one reason or another. The issue is one of ends and means. The structure by itself is in this case of little use – the question is whether its form in this particular setting warranted the effort which went into its creation. There were many doubters even amongst some who gave their all to make it happen.

At the time of the opening I was ill and could not attend. But I did go a few months later. After checking into our hotel my wife and I went down to Bennelong Point, the peninsula on which the Opera House has been built. It was evening; Sydney Harbour, always a majestic setting, had that special quality of water, sky, light and the reflection. The Opera House, which had been nothing but a building site for nearly fifteen years, was crowded with hundreds of people; performances were in full swing and the mass of people crowding the forecourt and

Sydney Opera House.

broadwalks were there just to enjoy themselves. A new great place had been created which gave pleasure to all who experienced it. And today, every day, throngs of people, not just theatre or concert goers, walk around the peninsula, stroll or sit and look at the harbour or have a drink in the open-air restaurant: it is a place to enjoy. And really that ultimately must be a testament to its success, a reference for future projects of its kind. So one of my reasons for its inclusion is that it is one of the very few, if not the only structure which has in my own experience made a major contribution to creating a place, a sense of space to be enjoyed by all who experience it.

My next choice leads me almost naturally from the Opera House. On my return from one of my early visits to Australia I was travelling with Jørn Utzon, the original architect whose competition design was the winning one. We had stopped in San Francisco to talk to T.Y. Lin about lightweight prestressed concrete which we were considering using on a rather larger scale than had been done before. (We eventually used concrete with normal aggregates.) We, in particular Utzon, were keen to look at one or two of Lloyd Wright's creations which he designed shortly before he died. So it was in the company of Utzon that I first set eyes on the Golden Gate Suspension Bridge. Walking down an ordinary street with Utzon was an interesting experience. The artist in him enabled him to see and articulate visual images and perceptions which one normally didn't notice, or at the very least made one see things in a totally different light. The Golden Gate Bridge would have bowled me over with or without Utzon. But having him around helped me absorb this marvellous structure on a totally different visual scale.

The Golden Gate at San Francisco is an extraordinary setting. The broad sweep of the Pacific on the one side, the hills, the fog rolling in, the stretch of water over a mile wide, all help to create a stage for an engineering masterpiece. But the story behind the project, the initiatives which helped to bring it about, the concept, the detailed design and of course the construction of the bridge is fascinating. (The stories behind many major engineering works are fascinating, even romantic and we should do more to bring them to the forefront of the media and hence the public.)

A larger than life Irish-American engineer, one Michael O'Shaughnessy, became the City engineer of San Francisco in 1912[2]. He was responsible for much of San Francisco's post-earthquake infrastructure, but it was he, probably more than anyone, who at the time took the initiatives which led to bridging the Golden Gate. That it would have been bridged sooner or later was inevitable, but the manner in which it came about was certainly the work of O'Shaughnessy. He discussed the crossing with many bridge builders and eventually, and rather surprisingly, turned to Josef Strauss, an engineer whose firm was based in Chicago, for the first serious examination of the problems for building a bridge over the Golden Gate. Surprising, because although Strauss had designed and indeed patented some bridges, they were modest structures of no great technical interest.

---

[2] *Spanning the Gate*. Stephen Cassady, Squarebooks, May 1987.

THE GOLDEN GATE BRIDGE.

Great bridge builders and designers like Roebling and Amman were there to be consulted, and Strauss's subsequent role is fascinating in that much of the technical work was clearly done by others more able and qualified than he to carry them out. Ellis, Strauss's chief engineer, appears to have been mainly responsible for the work. He was aided by a board of distinguished engineers including Amman and Moisseiff who advised and checked the work. I will return to this issue of attribution of great projects: it is a matter which in this day of media power has some significance.

The combination of setting and design make the Golden Gate bridge rather special. Many great suspension bridges have been built since, but none have surpassed its dramatic impact. The more I learnt subsequently about its genesis and the men who created it, the more my admiration for this unique structure has increased. Joseph Strauss, the ambitious entrepreneur-engineer who is usually attributed with the creation of the bridge, was apparently not enthusiastic to acknowledge the contribution of others, particularly his own colleagues, Ellis and Paine! The Golden Gate Bridge is also typically a product of the American democratic system, where participation at grass roots level in a major public undertaking is commonplace, but where, at the same time, opportunities for legal prevarication as well as more questionable practices are plentiful.

The towers are made up of bundled 42" square tubes dropped off to create a taper, with decorative brackets for the portal beams, which at the same time have a sensible structural function. These are particularly inventive. The whole bridge remains one of the most inspiring engineering works of this century.

My next choice is the Terminal Building at Stansted airport. The reality of the old cliché that 'it is better to travel hopefully than to arrive' has all but disappeared. Train and plane travelling is part of the packaging industry. Travelling by sea without a car has all but disappeared (with a car it has become a container within a container experience.) Our expectations now anticipate arrival before departure. It was not always so; the great railway stations – Paddington, St Pancras, King's Cross – are still evocative of the anticipation and excitement of travel.

Air travel, that dramatic twentieth century addition to man's travel experience, has not delivered in its arrival and departure spaces the kind of experiences one would expect from such a technologically contemporary mode of transport. Stansted airport in its design concept attempts to retain that connection between traveller and mode of transport which in all but the smallest and most primitive airports has nearly disappeared. The structure plays a significant, if not the key, role in achieving this objective. It is also a structure which is repetitive, always a good economic omen. It is also light and it is one of the few structures where the 'tree' concept has been followed without introducing a rash of complexities.

THE TERMINAL BUILDING AT STANSTED AIRPORT.

KING'S COLLEGE CHAPEL.

When I was a student, and even as a young graduate engineer, the very word 'vierendeel' evoked irrational prejudices – a Continental European development, heavy, expensive and in those pre-computer days, a computational hurdle of significant proportions. Isn't it strange how labels distort reality? It is just another structure, a frame more or less rigid depending on the designer's intentions. In the structure for Stansted, early schemes clearly illustrate some of the built-in assumptions. Removing the cross-bracing had negligible economic impact and was quite necessary for the sensible functioning of the building.

Stansted is on my list of eight because not only is the structure clear, light and appropriate, but the details satisfy. They are almost tactile: the human, the acceptable face of technology.

My last but one choice is King's College Chapel in Cambridge[3]. Work on King's College Chapel started in 1448. Henry VI in his famous 'Will and Intent' started what became a saga lasting more than 60 years. Its construction spanned a turbulent and formative period of English history. Four masons were responsible for its design and construction (masons now called architects but who could in many ways equally be called engineers). The last of these masons, Wastell, who took over the work in 1508, was faced with an unfinished and half-built shell. He completed it and designed and built a vault for the whole building which has turned out to be a *tour de force*, a high point in English Gothic design.

[2] *Cambridge Architecture*, T. Rawle, Trefoile Books, 1985; *Observation on the construction of the roof of King's College Chapel, Cambridge*, R.F. MacKenzie, John Weale, Architecture Library, 59 High Holborn 1840.

THE INTERIOR OF KING'S COLLEGE CHAPEL.

On detailed study it is clear that the chapel is not the work of one hand; there have been changes in design, which are thought to be a consequence of the different stylistic predilections or training of the four masons who were, at different times, in charge of the work. Yet despite these substantial detailed differences, particularly noticeable in the vault support system, the chapel is a unified whole. Many buildings of the time, the great churches in particular, often fail to be legible as a totality – something to do with scale and changes in style. King's College Chapel is a comprehensible whole, its scale slightly odd because it was originally meant to be part of a college cloister, but the impact of the vaulted roof is quite amazing. It has been described as the best-planned, best cut and best executed stone vault in England.

My eighth and last choice is the television and telecommunication tower, the 'Torre de Collserola' in Barcelona. The design of the tower was the subject of a design competition. The objective of this, the winning design, was to derive a tower from first principles to accommodate inevitable changes in telecommunication technology over the anticipated life of the tower: a tall order in the current climate, but nevertheless a desirable feature if achievable.

The structure was required to sustain lateral loads with minimal distortions to minimize 'downtime'. It was also to be sabotage resistant, an unfortunate necessity that has become a feature of late twentieth century design. But the most challenging aspect of the design was not to desecrate the skyline, for the site is situated on the top of a hill on the outskirts of Barcelona and if insensitively handled could become visually domineering. The structure was required to be 288 metres high and a conventional concrete tube or its equivalent would have meant a footprint dimension of at least 25 metres.

Torre de Collserola, Barcelona

The solution was to combine the lateral and vertical load systems and at the same time to minimize the use of materials and make the structure as transparent as possible. The full width of the tower is used by three primary trusses stabilized by three pairs of guys at the bottom. At the top, three single guys made from Kevlar, a non-conducting polymer, transparent to radio waves, connect the trusses to the central shaft. This central shaft, only 4.5 metres in diameter, supports the vertical load and contains the main telecommunication cable and power risers.

Looking at this selection of eight structures, I attempted to find reasons for choosing these rather than many others which were on my short-list which could be said to have equal or greater merit. I did find some common threads which I will try to articulate.

In each case there is a comprehensible, legible idea. Whether or not we respond more positively to something which is comprehensible is open to debate: certainly when that idea is associated with visual or spatial delight a positive response is more likely. The clarity of an underlying theme is appealing.

The Forth Bridge is the image of its famous human 'prototype' (see below). It is not difficult to develop an affinity to something which is so comprehensible. The clarity of Frankfurt, of Stansted, and the Barcelona tower is also direct. Sydney Opera House is another matter. Utzon, the original architect, entered the competition uncluttered by engineering advice. A cynic might say that this is obvious. There was no clarity in the shapes he was proposing, nor did he have a real under-

'WORKING MODEL' OF THE FORTH RAILWAY BRIDGE.

EARLY SKETCH OF SYDNEY OPERA HOUSE BY JØRN UTZON.

standing of how the structure of the roof could resist the forces to which it would be subjected. The development which took place after he had won the competition meandered through a series of geometric definitions: structure creating geometrical order has a unifying and aesthetic resonance. The final built spherical geometry gives a substantially changed silhouette from earlier schemes, particularly the free undefined shapes of the competition. Its visual order makes the structure crisper and more comprehensible.

King's College Chapel, despite taking 67 years to be built (Sydney Opera House took a mere 15), is an immediate visual entity. And even the West African mosques have a degree of coherence. So that the coherence of these eight projects has a seductive appeal.

All eight structures are totally integrated: they are part of the architecture and, in the case of the two bridges, they are architecture in themselves. But even here the multidisciplinary nature of design is an important factor. In the case of the Golden Gate Bridge, Morrow, an architect, was on Strauss's staff for many years. He was a Modernist, quite an accolade in 1930, if not now. He was very influential in the design of the towers. He felt that the portal bracing should diminish above the deck; he introduced fluted coverplates and helped to design the distinctive corner brackets, both a visual and structural improvement. Above all he chose the famous colour, 'international orange' as it was officially called.

The involvement of architects in the design of the Forth Bridge is not clear. Sir Benjamin Baker who, more than anyone, was responsible for the design, certainly articulated strongly the need for interprofessional collaboration. During the discussion of a fascinating paper on the 'Aesthetic Treatment of Bridge Structures' at the Institution of Civil Engineers in March 1901, he told a story against himself. 'At the early age of twenty-two he had thought he could do without architects and he had designed and carried out some very pretty work indeed. It had been so pretty that it had attracted the attention of the redoubtable Mr. Ruskin, who had mentioned it in one of his lectures. There had been columns and arches and scrolls in ironwork and Mr. Ruskin had said that he had seen it, and that it had

CONCORDE.

made him wish he had been born a blind fish in a Kentucky cave. Sir Benjamin thought that Mr. Ruskin had let him down very easily, because sometimes he could say nasty things!'[4]

The architects for the buildings are, of course, well known: Foster Associates for the sports stadium, Stansted terminal and the Barcelona tower, and Utzon for the Opera House. The four master masons who designed and built King's College Chapel were well-known in their day. The designers for the African mosque are not easily established: records here are not easily attainable. There remains, of course, the burning issue of who did what and what role the various participants played. I wish it became part of the public's conventional wisdom that while the artifacts we are discussing are not exactly designed by committee, they are nevertheless the products of many minds. For instance to try and establish who designed Concorde is impossible. It is the product of a series, a sequence of planes built and unbuilt, the endproduct of the creative efforts of many individuals. It is fashionable, particularly in this age of media hype, to concentrate on the cult of the personality. That the final result may well bear the stamp of one pair of guiding hands often obscures the involvement and the equally important contribution of many others.

Some time ago I was given a book called *Making Connections – Teaching and the Human Brain*.[5] It is a book difficult for non-specialists to understand, but a chapter

[4] *Proceedings of the Institution of Civil Engineers.* 1900–1901. Discussion of paper *The Aesthetic Treatment of Bridge Structures.* Joseph Husband.
[5] Renate Caine, Geoffrey Caine. ASCD (Association for Supervision of Curricular Development), Alexandria, Virginia.

headed 'The mystery of meaning' started with a quotation which I did understand. It goes as follows:

> *Two stonecutters … were engaged in a similar activity. Asked what they were doing, one answered, 'I'm squaring up this block of stone.' The other replied, 'I'm building a cathedral.' The first may have been underemployed; the second was not. Clearly what counts is not so much what a person does, but what he perceives he is doing it for.*

To create the kinds of structures I have chosen requires the talents and commitment of many 'stonecutters' and they must all feel involved in the whole, not just the part. That is the challenge in this increasingly complex world of ours if we are to satisfy our clients and continue to seek excellence in what we do.

I said earlier that I would specify the luxury item allowed, in addition to the Bible and the works of Shakespeare. A computer would be very tempting so that I could finally learn to master the thing, but whatever I learn will inevitably be old hat before long. But I decided on a piano to which I must add a 'do-it-yourself learning' kit. I have always regretted not having learnt to play and the desert island would seem to be the perfect opportunity. Indeed I wonder whether someone, someday, will make a connection between the structure of beautiful harmony and the harmony of a beautiful structure.

Acknowledgements

We record our thanks to Mrs Diana Plomley for consent to use in this chapter the format of *Desert Island Discs* created by her late husband, Mr Roy Plomley.

Thanks are due to the following for illustrations:
The Forth Railway Bridge (courtesy IMSI Masterclips and MasterPhotos Premium Image Collection, 75 Rowland Way, Novato Ca. 94945, USA).
A photograph of the model of the National Athletics Stadium, Frankfurt (photograph courtesy Foster Visualisation).
CAD image of the repetitive diamond-shaped unit of the National Athletics Stadium, Frankfurt (photograph courtesy Foster Visualisation).
The Great Mosque at Djenne (photograph courtesy Jean-Louis Bourgeois and Aperture Publications).
Sydney Opera House (photograph courtesy David Messent Photography).
The Golden Gate Bridge (courtesy IMSI Masterclips and MasterPhotos Premium Image Collection, 75 Rowland Way, Novato Ca. 94945, USA).
The Terminal Building at Stansted airport (photograph courtesy British Steel plc).
King's College Chapel (Photo © Woodmansterne).
The interior of King's College Chapel (Photo © Woodmansterne).
Working model' of the Forth Railway Bridge   (Photograph courtesy Institution of Civil Engineers).
Eraly sketch of Sydney Opera House by Jørn Utzon (Photograph courtesy Ove Arup & Partners).
Concorde (Photograph courtesy British Airways).

Sir Jack Zunz

BSc, FREng, FICE, FIStrucE, FCGI, Hon DEng, Hon DSc, Hon FRIBA

Sir Jack was born on Christmas Day, 1923. After interrupting his studies to serve with the South African Forces in Egypt and Italy in the Second World War, he graduated in civil engineering at the University of Witwatersrand in 1948. After two years with a consultant and a structural steel fabricator, he came to London to join Ove Arup in 1950. In 1954, he returned to South Africa where, together with Michael Lewis, he started a practice for Arups.

In 1962 Sir Jack returned to London as an Associate Partner and then from 1965 as a Senior Partner and he led the team which designed the roof of Sydney Opera House. Subseqently, he was responsible for many landmark projects, including the Standard Bank building in Johannesburg, Britannic House for BP, the Emley Moor transmission tower, Hongkong and Shanghai Bank and Stanstead Airport terminal building. He was chairman of Ove Arup and Partners from 1977 to 1984 and Co-chairman of Ove Arup Partnership, the whole Arup group, from 1984 to 1989. He helped the firm to develop technically, increase its geographical spread as well as create the framework to accommodate an increasing number of talented engineers.

Sir Jack was consultant to Arups from 1989 to 1995 and the first Chairman of the Ove Arup Foundation. Under his guidance the Foundation initiated multi-disciplinary post-graduate programmes at the University of Cambridge and the London School of Economics. He was a Fellow Commoner at Churchill College Cambridge 1967–68. He has lectured widely on his projects and related topics, particularly education, and is the author or co-author of many papers. He is the Chairman of the Trustees of the Architectural Association and was President of CIRIA from 1985 to 1998. Together with Sir Ove Arup he was awarded the Institution of Structural Engineers Silver Medal in 1969 and he received that Institution's Gold Medal in 1988.

Some features of a career relevant to the future aims for a young engineer

Tony Flint

*I*n my contribution to this volume I have chosen to review certain facets of my career which may be of interest and help to young engineers. Among the maxims to which I have held in my professional life, several have served me well. Those relevant to this contribution (and there are others) are:

- Grasp opportunities when they arise.
- Make the most of patronage and repay those who place confidence in you with dedicated service.
- Be prepared to tackle something beyond your experience.
- Learn from failures, both structural and in your own performance.
- Pursue excellence.

Looking back I find it remarkable how one thing has followed another in expanding series in my career as a result of those tenets. I will illustrate this by means of examples.

On graduating, my Professor at Imperial College, Sutton Pippard, encouraged me to join the Royal Aircraft Establishment at Farnborough to take part in research into the behaviour of aeronautical structures, although I had studied civil engineering. I found myself among engineers and scientists with numerous disciplines who had competence and flair in very diverse fields. I learnt how to measure strains in airframes (my introduction to the value of instrumentation) such as that for the revolutionary Brabazon, which was the unsuccessful forerunner of the modern airliners, and studied the aerodynamics of the flight of golf balls. (My boss was a keen golfer!)

The head of the structures department, Dr (later Sir) Alfred Pugsley, left to take up the Chair of civil engineering at Bristol University and set about to establish a postgraduate school. He recruited me as one of his first two PhD students and set me to study the lateral stability of beams, a topic which later opened surprising opportunities. Pugsley was a wonderful mentor with many interests, including those of the behaviour of suspension bridges and structural safety, and remained so throughout his life.

On completing my thesis I found there to be gratifying interest in my work. At the time, two new structural Design Codes, BS 153 for steel bridges and BS 449 for steel buildings, were being drafted and both needed clauses dealing with the buckling of beams and plate girders. Following a daunting cross-examination by Dr. Oleg Kerensky of Freeman Fox & Partners, a giant in the bridge engineering field, I found myself on the committee for BS 153. I have remained on the Code Committee ever since, later being entrusted to chair the sub-committees to develop the steelwork parts of BS 5400. One thing led to another.

While at Bristol the condition of Clifton suspension bridge came under review and Pugsley's expertise was drawn upon. I was deputed by him to undertake field measurements of strains in its cross-girders, and tensions in its hangers, and to compare the observed behaviour with theory. Thus was I drawn into the field of suspension bridges which later culminated in my appointment to assess the adequacy of the designs of a number of the world's largest bridges and to design the strengthening of the first Severn suspension bridge. In between I had the privilege of contact with Sir Gilbert Roberts, who was responsible for the then revolutionary design of Severn bridge, Sir Ralph Freeman, and eminent bridge builders such as Hubert Shirley Smith, via my activities associated with the BS 153 Committee. Throughout I was enthused and taxed by my encounters with Oleg Kerensky and with the then Chief Engineer of the Ministry of Works, Harold Gardiner.

The drafting of the new bridge Code began in 1969. It was to be the first limit-state partial factor code for bridges in the world, and posed many new philosophic and practical problems. At the time when it began there was a new trend towards the use of steel box girders for the decks of long-span bridges. There had been no experience in this form of construction since the days of Stephenson's wrought-iron Britannia bridge, and the need for research into the behaviour of the same form in welded steel construction became apparent to my sub-committee. (Stephenson had enlisted the help of Fairbairn and Eaton Hodgkinson to undertake the necessary research for his bridge.) The only codified information available for design was that for steel plate girders in BS 153. Despite endeavours to obtain funds for such research, we failed to receive any funding, partly on the basis of the contention that the requisite information was available from research into the behaviour of air frames. It was therefore ironical that before the Code had been drafted four major box girder bridges collapsed during construction for a variety of reasons. This opened the floodgates of funding of research after the horse had bolted! Learn from this.

A Government Committee of Inquiry into the collapses at Milford Haven and Melbourne was set up under the chairmanship of Sir Alec Merrison, on which I served. Jointly with Professor Michael Horne, a brilliant engineer with outstanding knowledge of buckling problems, I participated in the unduly rapid drafting of the 'Merrison' rules for the assessment and design of box girder bridges. The entire national expertise in testing and research into the relevant potential problems was mobilized and some very valuable results were obtained which served to put the United Kingdom into the forefront of knowledge in this field, where it remains today.

Learning from failures is ultimately the way ahead in the uncertain world of engineering. In the undertaking of this task, I had need, as I still do, to revert to my training under Pugsley, and to my knowledge of bridge design learned from Roberts, Freeman and Kerensky and their team of designers.

Inevitably the law intervened in the free transfer of experience from these events. I was engaged by the contractors for Milford Haven bridge, not only to advise on the completion of erection after the collapse, but also to provide evidence for potential litigation. The outcome of many investigations into the causes of collapse were, as a result of being *sub judice*, never published. In consequence the profession was deprived of much useful knowledge. This has been my experience in a number of other instances.

Following another of the paths of my career, I moved back to the patronage of Professor Pippard in Imperial College and, while there, I was invited to study the behaviour of the towers of the new Forth bridge, which had vibrated under wind during erection. I was subsequently asked by British Insulated Callender Construction Company whether I could advise on the aerodynamic stability of the Crystal Palace television tower. Although I had not previously been engaged in the field of vortex-excited vibrations of structures, I felt able to undertake the relevant analyses using the then current knowledge. This led to some exciting experimentation firing rockets from the top of the tower and a long-lasting collaboration not only with BICC but also a most valued friendship with Kit Scruton of the National Physical Laboratory who was the leader in the field of aerodynamic instability, not only for towers but also for major bridges.

My patronage by BICC led me increasingly into the design of towers and, subsequently, guyed masts. In the 1960s a new generation of 400 kV power transmission lines throughout the UK were to be constructed, entailing the design of towers taller and more robust than their predecessors. It was the practice to test prototypes in a special facility at Cheddar, and it was early discovered that designs by various contractors failed before reaching their design loading. Engaged to investigate the causes, I discovered that the problem arose from moving into a previously uncharted situation for which empirical rules for the design of bracing broke down. The legs of the new towers were not only larger than before, but also had greater bending stiffness. As a result the strains in the legs gave rise to forces in the bracing which were not accounted for, and this caused the premature failures. The investigations led to the development of non-linear methods of analysis, advanced in their time, which promulgated new analytical techniques in the early days of electronic computing. Again, knowledge was gained from failures (not catastrophic) and the problem was overcome.

Further knowledge was gained as a result of my appointment with Preece, Cardew & Ryder and Fairhurst & Partners to investigate the cause of the collapse, in a gale, of the towers supporting the transmission crossing over the river Clyde. This followed an earlier failure of similar towers on the Tees. Again the causes were attributed to extrapolating design practice beyond the range of previous experience, as has frequently been the cause of structural disasters. No lives were lost and I was engaged as advisor for the reconstruction, which benefited from the knowledge gained.

In a similar vein, I investigated the cause of collapse during erection of a radio mast at Waltham which resulted from acceleration of wind speed over the hill which, at the time, was not properly accounted for in design procedures, but which now is.

Perhaps the most complex failure which I investigated was that of the guyed transmitting mast at Emley Moor. In that instance I was engaged as one of the three-man team of enquiry for the Independent Broadcasting Authority. The mast collapsed in thick fog after the formation of ice on the stays. No-one witnessed the collapse. The evidence of the icing melted on hitting the ground. My previous experience in vortex-excited vibrations came into play, but the phenomena of icing were unexplored and required study from scratch. The resulting legal action brought me again into the field of litigation, and took 13 weeks of High Court action, followed by appeal and, subsequently, further appeal to the House of Lords. The initial verdict in favour of the Authority was upheld by their Lordships, albeit for the wrong technical reasons.

After early years in academic life I had a yearning to design something. Again, kindly patronage, this time from John Henderson, an eminent lecturer in concrete technology, steered me on a new course on which I embarked with enthusiasm. Henderson introduced me to the eminent architect, James Cubitt, who was seeking an engineer to undertake the structural design of his buildings in his offices. Despite my paucity of experience and a largely irrelevant CV, Cubitt entrusted me with the task, and there followed a long harmonious collaboration in creating designs for universities, hotels, factories and schools in the UK and Africa. Cubitt also agreed to my undertaking consultancy for others in parallel and Henderson again promoted a meeting with Peter (Jo) Chamberlin of Chamberlin, Powell & Bon, who placed his confidence in me to design the structures for his design of a school in Southwark. Chamberlin had a policy of trying something new every day, and this proved to be the most stimulating tenet. This was an era when concrete shell structures were becoming fashionable, and he decided to roof the assembly hall of the school with hyperbolic paraboloid shells. The budget was tight and such shells constructed in conventional ways were known to be expensive. A chance meeting with a Mr Hart of the Gunite Construction Company led to the idea of constructing the roof by spraying the concrete from within onto insulation supported on a cat's cradle of prestressing wires suspended from permanent steel framing. Not relying on my calculations (there were no finite element programs in those days, and the proposed technique had never been tried before), Mr. Hart made a model constructed in the same way which survived all the sandbags which we could load onto it. Thus began a lasting involvement with concrete shells, including the first ferrocement dome for New Hall, Cambridge (only 20 mm thick,) the gunited ventilation towers for Blackwall tunnel, the gunited conoidal roof to Birmingham University sports hall and culminating in the 81 petals of the Baha'i House of Worship in New Delhi. By then, computer power had developed to an extent such that we could model the most complex geometric forms and calculate stresses with greater confidence. It was, however, noteworthy that the latter highly complex structure was built by hundreds of Indian workpeople using basic skills.

A further venture into the unknown resulted from the problems encountered in integrating building services into a structural design. Chamberlin decided to entrust my firm to set up a co-ordinated team to undertake the mechanical and electrical services design for his buildings, although we had no track record in this. After a trial run for Cheltenham Grammar School, we were entrusted to provide the combined service for the whole of the post-war development of the University of Leeds. This experience, and that obtained with Cubitt in the design of the structure for the theatre for University College, London, led to the appointment to assist Sir Denys Lasdun in the design of the structure and services for the National Theatre. The Lasdun concept for this most complex building demanded complete integration of services of very diverse nature into the structural scheme. It was a taxing experience to achieve this, which I believe we did.

Reverting to the threads of my history, the advent of the limit-state partial factor structural design codes entailed consideration of the values of the factors to be applied in them to loading and resistance. A committee of the Institution of Structural Engineers, under Pugsley's chairmanship, had proposed a subjective basis which might be used to derive the factors. By 1975 the science of reliability theory had developed to such an extent that it offered a more rational basis, and in that year, with the support of Kerensky and others, the Construction Industry Research and Information Association (CIRIA) sponsored me and a group of experts to prepare a report on the rationalization of safety factors for design. Our report, built on the work of Pugsley, proposed procedures for calibration and outlined a possible methodology which appears to have been subsequently embraced worldwide. Although we were entrusted to provide the factors for BS 5400:Part 3, there was a subsequent loss of interest in the UK in the report and the applications of reliability theory which has only recently been revived by the Highways Agency. The message for young engineers must be to press for support to build on UK expertise.

One of the conclusions of the CIRIA work was that there should be a balance between the acceptable risk of loss of life as result of a structural failure and the cost of preventing such loss. This balance appears to have not been properly assessed in the imposition of recent legislation. As a result, disproportionate costs are being imposed on society in the name of 'safety'. Pugsley was concerned with the balance between the reliability of aircraft and their combat effectiveness. Today we need to carefully consider the cost/benefit analysis of any design safety provisions, and engineers should question politically imposed constraints on design or assessment which may be emotional rather than rational.

Another thread in the weft of my career has been woven into the tapestry, first formed in Farnborough and on Clifton bridge. In recent years I have been entrusted by the Hong Kong Highways Department to develop a system of structural health monitoring for three major bridges. This has resulted in instrumentation of the bridges of the most advanced scope and a masterplan for reviewing the reliability of the bridges into the future. One thing leads to another.

The tenet of the pursuit of excellence is, in my opinion, vital to the career of an engineer. In all of the above examples I have held that I must provide a service expected of me by those who have entrusted me to serve them and their client. Of course there have been pressures of time and the need for response by others who may have been on the critical path. It has been up to me to deal with these if I had the ultimate responsibility for progress. This has not always won friends, but adherence to programmes is an essential part of engineering life.

It has always been natural to me to try to deliver what was required on time. There was no need for the bureaucratic processes of Quality Assurance when I entered my consultancy practice. It was natural to check for the accuracy of design assumptions and calculations and to test for uncertainties. There is no substitute for self-checking, but the advent of the computerized age poses new problems.

The aim of this contribution has to be, by examples, to encourage the next generation of engineers in the way in which they may continue to lead the world in their art and science and to serve society in a professional way. Although changes in the way of working are inevitable, and desirable in some cases, I believe that the maxims which I have adhered to should survive, despite political and other pressures. Integrity is all.

Dr. A.R. Flint

BSc, PhD, FICE, FIStructE, FREng

After early training on the Great Western Railway, graduated from Imperial College and joined the Royal Aircraft Establishment, Farnborough. He then undertook structural research for a PhD at Bristol University, being subsequently appointed as Research Fellow until 1953. He was appointed Lecturer in Civil Engineering at Imperial College.

In 1956 set up in private practice as a structural engineer and joined J.A. Neill in 1958 to found Flint & Neill. In the same year he was appointed London University Reader in Structural Steelwork, which he remained until 1972.

He was in charge of the design of schools, university buildings, theatres and special structures including the National Theatre and the Baha'i Temple in New Delhi. He undertook special studies and designs for tall towers and masts and specialized in structural aerodynamics. He served on the committee which drafted BS 153 for steel bridges and was later appointed to chair the subcommittees for BS 5400 drafting the steel sections of the code. Following service on the Merrison Committee of Inquiry into box girder bridge failures, he specialized in the assessment and design of strengthening of major bridges.

273

The holistic approach

Sir Alan Cockshaw

Hulme regeneration

In my early career I was concerned with the design and then the construction of new roads, bridges, water schemes, power projects and industrial and commercial developments. I thought then we were delivering total solutions but, of course, in fact we were only concerned with providing the technical and physical solutions to someone else's prescribed need or opportunity.

Surprisingly, it was an increasing involvement in opencast mining which began to stimulate a compelling interest in a broader, more holistic approach. In opencast mining we often started with greenfield sites and sometimes what are now described as brownfield sites. In each case for some years our efforts were significant 'blots' on the landscape, until the process of reinstatement began. At that time, reinstatement was simply about refilling the voids to the original ground profile before replacing the subsoil and topsoil. This was the approach on the very first site I worked on more than 40 years ago. However in later years we began to develop broader master plans to create a different topography, a new shape.

We provided lakes and stocked them with fish, jetties were built, even more trees were planted and progressively we began to create new areas for future development. Concepts were thus developed between our client, the local community, landowners and ourselves. It was, after all, easy to recognize the potential for added value in land restored in this way, especially when originally the land was partially derelict.

In a modest way, holistic solutions were achieved through the joint efforts of planners, engineers, landscape architects and a whole range of professionals working together to achieve excellence. It was the realization of what could be done in such projects that opened up the prospect of an even broader view of existing mining areas and the opportunity to become more involved in planning comprehensive solutions to some difficult and often disadvantaged areas, frequently adjoining urban or semi-urban areas.

This sort of long-term comprehensive planning of new sites began to allow, for example, the safe disposal firstly of routine domestic waste and later of more

HULME
REGENERATION.

difficult industrial and toxic wastes in carefully designed, safe 'coffin' areas. These later initiatives allowed land left unsafe and derelict for years to be restored to the considerable benefit of the whole environment around it. These were usually relatively large areas of land and therefore the potential for adding value was substantial if such long-term comprehensive planning and high quality engineering solutions were considered.

This kind of overall approach clearly offered and continues to offer significant sustainable long-term solutions, not only in former mining areas but in many former major industrial locations. Such solutions are socially desirable and environmentally friendly and create the potential for added-value and can be unashamedly profitable for those involved in their execution.

Partnerships are central to all of them, between major landowners, local authorities and indeed central government, and each scheme requires that same combination of high-quality planners, architects and engineers and the broad range of built-environment expertise.

Not surprisingly, therefore, when my colleagues and I were asked by the government and the City of Manchester to become involved in the regeneration of Hulme, we agreed to take part, only if:

- both central and local government were willing to take a holistic view of the entire area, not just the specific areas that were identified as the problem;
- government would integrate the spending of its different departments rather than have them competing with each other and the City would do the same;
- finally, we would have a dedicated private sector/public sector team charged with implementing a new vision for the area.

We were able to retain the services of some genuinely world-class planners to work with the key employers in the area and more importantly the people of the area to create a new vision and then to implement it. Individual neighbourhoods had separate planning and architectural teams working with the community to create the new vision. Hulme was an area of singular deprivation and lack of opportunity, where no-one wanted to live and no-one could work. Seven years later we have a new community with new jobs and an exciting future. It worked because of a partnership between the private sector and the public sector, between planners, architects and engineers, working together to deliver a total solution, a holistic solution. Sustainability in such schemes is just as fundamental as the quality of the original vision. The creation of opportunities for the people of the area is the vital factor in the sustainability of the solution and has been very much at the heart of all our efforts.

More recently Manchester Millennium was charged by the government and the City with the task of rebuilding the centre of Manchester after the IRA bomb a few years ago. We have adopted the same principles again.

All of the main building owners (whether their buildings were damaged or not) are part of the process of promoting the vision they have created in partnership with the City Council and the government. After all they are together providing the capital and expect to get a return on their investment. Again, we have world-class planners and architects, working hand in hand with engineers of many disciplines. We have a clear, shared and sustainable vision. Concurrently we are studying comprehensively the entire way in which people move around the city and are developing a comprehensive integrated transport plan. This marriage of land use and transport planning is clearly essential in any urban area.

This is perhaps my best and truly most holistic example with all the professionals working together to achieve some remarkable solutions. The bomb did just over £100 million of damage but the projects now being implemented will result in a total investment of almost £1 billion.

The benefits of the integration of all the professionals in specific projects to deliver total solutions are very substantial indeed, and the compulsion to achieve the advantages created is irreversible. The opportunity to incorporate this approach to individual projects within the holistic development of a new vision for any area offers exciting challenges. It certainly works everywhere I have seen it applied, from a modest open-cast coal site to a city centre.

Brenig Dam

I first became involved in Brenig Dam in December 1974. As a recently promoted, most junior director of a division of Fairclough Civil Engineering, I was given responsibility for the completion of the dam following the acquisition of Sir Lindsay Parkinson by our parent company, Fairclough Construction Group.

At that time the contract was about a year late, the losses incurred were very substantial indeed and it was likely that the client would shortly apply something of the order of £1 million in liquidated damages. The consulting engineer and the main contractor were at odds with each other; the subcontractors and the main contractor could not communicate effectively and the client, not unusually at that time, was very remote from the issues of the project.

In the end, we did something over two years' work in a year, finished the project on time to an excellent quality and finally had a very satisfied client. We, as main contractor, together with our sub-contractors, all felt we had been dealt with fairly.

Perhaps more than most projects in my career, this was one where the client, the consulting engineer, the main contractor, the principal sub-contractors and indeed at times the suppliers were about as isolated as they could ever become. The challenge was to find a way to focus all their minds towards the simple objective of working together to complete the job as quickly and economically as possible, yet at all times sustaining quality at every level.

When I first visited the site, foundation preparation and grouting were not proceeding quickly because the sub-contractor had the clear perception that the earthworks sub-contractor did not need to proceed at a pace that could ever catch them up. The processing of filter materials was consistent with the progress that had so far been made in dam construction and certainly not consistent with the real progress required. Consequently the plant resources for the main embankment were not capable of producing the outputs required for completion.

There were, of course, some very fundamental contractual disputes relating to the specification of the core, the filter materials and the laboratory control procedures. There was a disconnection between the technical arguments, the contractual arguments and the aims and objectives of all the participants. The solution was quite unsurprising: get the job done by focusing on the programmed outputs for each activity to ensure completion by the agreed date.

This clearly meant that grouting and key trench preparation were the key activities and were given precedence over all others, resulting in twenty-four hours a day working to rapidly unlock very substantial areas for embankment construction. Production of filters required 'round the clock' working to meet the intended programme for the construction of the core and the shoulders of the dam. The testing regime for all this required to be formalized and agreed with a joint programme of testing and monitoring.

There were very serious financial commitments to be made by the main contractor and the two principal sub-contractors and the development of confidence between those participants and the consulting engineer was vital. Happily that trust was developed. Confidence between each of the parties that they would achieve the objectives that we had jointly set grew so that production levels began to exceed our most optimistic target. It became a true partnership and together we achieved productivity and quality that nobody would have believed possible just a few months earlier.

Of course there were many arguments before the job was finally settled financially, but in today's terms it was done very quickly, indeed shortly after its opening by the Prince of Wales on 21st December, 1976. The project was blessed by having an 'Engineer' to the contract who really understood his role and responsibilities and who presided over all of the contract finalization. He cut through the parochial technical and commercial arguments that had so bedevilled the contract in its earlier years.

A satisfied client is, of course, always the key to any project and it is perhaps a pity that the client had not been involved much earlier. In the end it was about people – in the trust and confidence between the Engineer and the directors responsible for the main contractor and the sub-contractors and the trust and respect that they grew to have for each other. One or two sadly are no longer with us but the remainder, having gone through the trauma, remain friends to this day.

All of the lessons have lasted throughout my career.

BRENIG DAM.

Any contract where the participants retreat into their corner and do not face the challenges together is doomed to failure. It is vital that senior people in a company keep an eye on relationships on a site in order to ensure that they change them if they ever become a problem sufficient to prevent people working together as an effective team. That applies just as much to the client as it does to the contractor. There is no excuse for the key players in a contract keeping themselves isolated or indeed perhaps insulated from it.

If you don't do a good job on time, you haven't got much chance of making a profit.

Sir Alan Cockshaw

BSc, HonDEng, HonDSc, FREng, FICE, FIHT

Sir Alan Cockshaw, President of the Institution of Civil Engineers 1997-8, was born in July, 1937. After graduating in civil engineering from Leeds University, he went on to have a distinguished career in engineering and construction, culminating in the chairmanship of AMEC, plc.

Following an early career in both the public and private sectors, Sir Alan joined Fairclough Civil Engineering in 1973; he became its Chief Executive in 1978 and a member of the main board of Fairclough Construction Group in 1981. After the acquisition by Fairclough of the Press Group in 1982 and the creation of the AMEC Group, Sir Alan became Group Chief Executive of AMEC plc in 1984 and Chairman of AMEC plc in 1988.

Under his leadership, AMEC grew to become one of the largest engineering and construction companies in Europe, with annual revenues in excess of £2.7 billion per annum and more than 25 000 employees worldwide. Sir Alan was primarily responsible for the development of its international activities from the United States, through mainland Europe into the Middle East, Africa and South East Asia, where its multi-discipline design and management resources are in increasing demand. He retired as Chairman of AMEC in July, 1997.

Sir Alan is Chairman of Manchester Millennium Ltd and Chairman of the Roxboro Group plc. He is a Director of the New Millennium Experience Company Ltd in London, and Life President of the North West Business Leadership Team. His former public positions included Chairman of the Oil and Gas Projects and Supplies Office within the Department of Trade and Industry, which advises government on issues concerning the UK's oil and gas industry, and Chairman of the Overseas Project Board. He became Chairman of English Partnerships and the Commission for New Towns in October 1998.